SpringerBriefs in Molecular Science

Biobased Polymers

Series editor

Patrick Navard, CNRS/Mines ParisTech, Sophia Antipolis, France

Published under the auspices of EPNOE*Springerbriefs in Biobased polymers covers all aspects of biobased polymer science, from the basis of this field starting from the living species in which they are synthetized (such as genetics, agronomy, plant biology) to the many applications they are used in (such as food, feed, engineering, construction, health, …) through to isolation and characterization, biosynthesis, biodegradation, chemical modifications, physical, chemical, mechanical and structural characterizations or biomimetic applications. All biobased polymers in all application sectors are welcome, either those produced in living species (like polysaccharides, proteins, lignin, …) or those that are rebuilt by chemists as in the case of many bioplastics.

Under the editorship of Patrick Navard and a panel of experts, the series will include contributions from many of the world's most authoritative biobased polymer scientists and professionals. Readers will gain an understanding of how given biobased polymers are made and what they can be used for. They will also be able to widen their knowledge and find new opportunities due to the multidisciplinary contributions.

This series is aimed at advanced undergraduates, academic and industrial researchers and professionals studying or using biobased polymers. Each brief will bear a general introduction enabling any reader to understand its topic.

*EPNOE The European Polysaccharide Network of Excellence (www.epnoe.eu) is a research and education network connecting academic, research institutions and companies focusing on polysaccharides and polysaccharide-related research and business.

More information about this series at http://www.springer.com/series/15056

Morgan Chabannes · Eric Garcia-Diaz
Laurent Clerc · Jean-Charles Bénézet
Frédéric Becquart

Lime Hemp and Rice Husk-Based Concretes for Building Envelopes

Springer

Morgan Chabannes
LGCgE-GCE
IMT Lille Douai
Douai Cedex
France

and

Université de Lille
Lille
France

Eric Garcia-Diaz
C2MA
IMT Mines Alès
Alès Cedex
France

Laurent Clerc
C2MA
IMT Mines Alès
Alès Cedex
France

Jean-Charles Bénézet
C2MA
IMT Mines Alès
Alès Cedex
France

Frédéric Becquart
LGCgE-GCE
IMT Lille Douai
Douai Cedex
France

and

Université de Lille
Lille
France

ISSN 2191-5407   ISSN 2191-5415   (electronic)
SpringerBriefs in Molecular Science
ISSN 2510-3407   ISSN 2510-3415   (electronic)
Biobased Polymers
ISBN 978-3-319-67659-3   ISBN 978-3-319-67660-9   (eBook)
https://doi.org/10.1007/978-3-319-67660-9

Library of Congress Control Number: 2017952907

Printed on acid-free paper

This Springer imprint is published by Springer Nature
The registered company is Springer International Publishing AG
The registered company address is: Gewerbestrasse 11, 6330 Cham, Switzerland

Contents

Nomenclature

LHC	Lime and hemp concrete
LRC	Lime and Rice husk concrete
ρ_B	Bulk density of plant aggregates
ρ_A	Apparent density of a particle
ρ_T	True density of the solid phase
η_O	Open porosity in the particle
η_I	Intergranular porosity in bulk aggregates
η_T	Total porosity in bulk aggregates
C, S, H, A	CaO, SiO_2, H_2O, Al_2O_3 (cement chemist notation)
SEM	Scanning electron microscopy
BSE-SEM	Back-scattered scanning electron microscopy
TGA	Thermogravimetric analysis
XRD	X-ray diffraction
A, B and W	Aggregate, binder and water contents
B/A	Binder-on-aggregates mass ratio
W/B	Water-on-binder mass ratio
W_P	Prewetting water
W_M	Mixing water
ISC	Indoor standard conditions
OC	Outdoor exposure conditions
ACC	Accelerating carbonation curing
MC	Moist curing
TA	Thermal activation
CS	Compressive strength
E_C	Tangent modulus on the loading cycle
η_{IP}	Intergranular porosity within the hardened concrete
η_{TOT}	Total porosity within the hardened concrete
λ_P	Thermal conductivity (flow parallel to compaction axis)
λ_O	Thermal conductivity (flow orthogonal to compaction axis)
MT	Manual tamping

VC	Vibro-compaction
ROC	Rate of carbonation
d,m	Days, months
ITZ	Interfacial transition zone
p'_0	Initial effective confining pressure
q	Deviatoric stress
σ'_m	Mean effective pressure
M'	Stress ratio
φ_P	Peak friction angle
C	Cohesion
FM	Failure mode

Chapter 1
Introduction

According to the International Energy Agency (IEA), the building sector accounts for one-third of final energy consumption and global carbon emissions in the world [1]. In Europe, most countries adopted their own thermal regulations after the first oil crisis in the 1970s in order to limit heat loss in buildings. In an effort to reduce energy consumption for the heating, some buildings were sealed too tightly without adequate ventilation, leading to poor indoor air quality. Furthermore, the thermal comfort in summer has been gradually considered through the different amendments of thermal regulations but it was largely ignored until the 1990s. The energy demand for air conditioning has increased in southern countries but not exclusively. It also applies to countries with a cold climate. Overheating in summer or even in the mid-season is frequently noted in Germany or Nordic countries where buildings are designed with high levels of thermal insulation, low permeability and solar heat gain through the glazing. Air conditioning has become relatively common in tertiary buildings even though it strongly affects the climate [2]. Within the framework of the Kyoto protocol, the European Union (EU) adopted a directive for the energy efficiency of buildings (Energy Performance of Building Directive known as EPBD) in 2002. It was revised in 2010 in order to provide harmonized methods for calculating the energy performance of buildings in thermal regulations, taking greater account of heating and cooling installations [3]. The French Thermal Regulation has been developed and strengthened several times. In 2012, the aim of the latest version was to achieve a decrease of 38% in the energy consumption of residential and tertiary buildings by 2020 compared to 2008 and a fourfold reduction of greenhouse gas emissions by 2050 in comparison to the level of emissions in 1990 [4]. Half of the building stock was built before 1970 and thus without any thermal insulation. Furthermore, over the following decades, heat insulation systems and design methods were not necessarily appropriate as briefly mentioned above (overheating, tight houses without efficient ventilation, lack of breathability, wall condensation, excessive use of air conditioning). Due to the low

© The Author(s) 2018   1
M. Chabannes et al., *Lime Hemp and Rice Husk-Based Concretes for Building Envelopes*, Biobased Polymers, https://doi.org/10.1007/978-3-319-67660-9_1

renewal rate of the building stock, the energy retrofit of existing buildings is absolutely fundamental to achieve the targets in terms of environmental impact. The construction industry is able to provide a significant potential for the reduction of greenhouse gas emissions. The energy efficiency of buildings during their operational phase tends to improve over time as a result of increasingly advanced insulating materials. Nevertheless, it is essential to pay close attention to the carbon footprint of selected materials. The environmental impact of building materials is not taken into account in European and French standards even though the embodied energy of materials is a key factor of the whole life cycle of buildings (Life Cycle Analysis approach). Conventional construction systems used for residential buildings mostly combine an insulating layer with a load bearing structure (concrete blocks). Mineral wools and polystyrene cover almost the whole market of insulating materials despite their high carbon footprint [5]. In order to keep buildings free from the risk of water condensation in traditional envelopes using mineral wools and plasterboard, self-insulating blocks like autoclaved aerated concrete or lightweight clay bricks have expanded in recent years. These load-bearing blocks have attractive hygrothermal properties [6] but they use non-renewable resources and their carbon footprint remains high (especially that of fired-clay bricks) [5]. The last two decades have witnessed the emergence of bio-based building materials mixing plant-derived aggregates with mineral binders. This return to old building methods is arousing great interest. In this field, hemp concrete has been well researched. It is designed by mixing hemp shives (the woody part of hemp stems) with lime or other binders. Crop residues are renewable resources and their use does not harm the environment. The carbon footprint of hemp-based concretes was found to be negative due to carbon sequestration during hemp growth and lime carbonation during the hardening of the concrete [5, 7]. Hemp concretes are manufactured with a high volume fraction of shives providing an important porosity to the hardened material. As a result, they show low thermal conductivity and good ability to buffer temperature and humidity variations. Hemp concrete prevents condensation, allows buildings to breathe (air-tight but water permeable). Thus, this bio-based concrete reduces heating and air conditioning needs while ensuring good indoor thermal comfort [8–10].

It is obvious that the diversification of renewable and easily available plant resources promotes and develops biomass-based construction contributing to carbon storage and sequestration. Rice is the first cereal in the world for human food. It is locally grown in the South of France. The outer covering of rice grains (called rice husk) is often considered as a waste material. This crop residue causes critical problems in rice growing areas given that high volumes are generated and not used in a beneficial way. The recovery of whole rice husk without any burning or grinding to design a lightweight insulating bio-based concrete was almost unexplored before. The new plant aggregate is mixed with lime-based binders and macroscopic properties of the innovative rice husk concrete are compared with those of hemp concrete.

Only the lime-based binders will be considered in the book. Their carbon footprint is much more favorable than that of Portland cement [11]. In addition, the water-vapor permeability, the low density (especially that of hydrated lime), the hardening through carbonation and the ductile behavior of lime binders are considerable assets for the mixing with hygroscopic and deformable plant aggregates.

Mechanical properties of hemp concretes depend on many factors such as casting process, binder content, type and size distribution of aggregates, curing conditions and age. In most cases, hemp concrete for a wall application is manually tamped into a wooden framework (cast on the building site) and the on-site implementation is conducted in accordance with professional guidelines [12]. In this case, the mechanical performances of the plant-based concrete are very low. The main weakness of hemp concretes using lime as binder is the long time they require to cure when cast on-site. However, the hardening through the carbonation process is known to provide additional strength to the material over time [13]. In addition, very little prior research has been done about the effect of curing conditions (relative humidity, temperature or even CO_2 content) on binder hardening and strength development of plant-based concretes. Using precast blocks is another option for the construction of walls with plant-based concretes. This method opens up interesting ways of improving the early age strength. Optimal curing conditions in order to accelerate the hardening of plant-based concretes should be subject to research as part of precast industry. In addition, some authors [14–16] studied the effect of high compaction of freshly-mixed hemp concrete under static loading. After hardening, hemp concrete shows significant increase in compressive strength and ductility. Whether they are cast in situ or in the form of precast blocks, plant-based concretes are only considered as insulating materials. The structural design practice of wood frame walls associated with hemp concrete does not assume any contribution of the plant-based material whereas it may contribute to the racking strength of walls. More knowledge about the shear behavior of plant-based concrete is needed to optimize the structural design. One can also identify insufficient hindsight towards the durability of this kind of material. Some authors began to turn their attention to this research focus [17].

The book provides a three-step outline:

- The first part proceeds with physical and chemical characterization of plant-derived aggregates (hemp shiv and rice husk).
- The second part mainly deals with the effect of curing conditions on hardening mechanisms and strength development of lime-based binders.
- The last and most significant chapter addresses mix design and hardened-state properties (porosities, thermal conductivity, compressive strength and even shear strength) of hemp and rice husk-based concretes. A large part is devoted to the influence of curing conditions on the strength development of manually tamped plant-based concretes. In addition, the results from triaxial compression of vibro-compacted plant-based concretes are presented.

References

1. IEA, *Transition to Sustainable Buildings: Strategies and Opportunities to 2050* (IEA, 2013)
2. U. (UBA), Building air conditioning in Germany 2015. Available: http://www.umweltbunde samt.de/en/topics/economics-consumption/products/fluorinated-greenhouse-gases-fully-halo genated-cfcs/application-domains-emission-reduction/building-air-conditioning. Accessed: 01 Jan 2017
3. European Parliament, Directive 2010/31/EU on the Energy Performance of Buildings 2010. Available: http://ec.europa.eu/energy/en/topics/energy-efficiency/buildings
4. S. Amziane, L. Arnaud, *Les bétons de granulats d'origine végétale Application au béton de chanvre* (Lavoisier., France, 2013)
5. M.P. Boutin, C. Flamin, S. Quinton, G. Gosse, Analyse du cycle de vie d'un mur en béton de chanvre banché sur ossature bois (2005)
6. A. Evrard, Transient hygrothermal behaviour of Lime-Hemp Materials, Ph.D. Thesis, Catholic (University of Louvain, Belgium, 2008), p. 140
7. A. Arrigoni, R. Pelosato, P. Melià, G. Ruggieri, S. Sabbadini, G. Dotelli, Life cycle assessment of natural building materials: The role of carbonation, mixture components and transport in the environmental impacts of hempcrete blocks. J. Clean. Prod. **149**, 1051–1061 (2017)
8. F. Collet, S. Pretot, Thermal conductivity of hemp concretes: variation with formulation, density and water content. Constr. Build. Mater. **65**, 612–619 (2014)
9. F. Collet, J. Chamoin, S. Pretot, C. Lanos, Comparison of the hygric behaviour of three hemp concretes. Energy Build. **62**, 294–303 (2013)
10. A.-D. Tran Le, Etude des transferts hygrothermiques dans le béton de chanvre et leur application au bâtiment, PhD Thesis, (Reims Champagne-Ardennes University, France, 2010) p. 209
11. M. Chabannes, Formulation et étude des propriétés mécaniques d'agrobétons légers isolants à base de balles de riz et de chènevotte pour l'éco-construction, (University of Montpellier, France, 2015), p. 215
12. C.en Chanvre, *Constuire en Chanvre. Règles professionnelles d'éxécution*, (SEBTP. Société d'Édition du Bâtiment et des Travaux Publics, 2012)
13. L. Arnaud, E. Gourlay, Experimental study of parameters influencing mechanical properties of hemp concretes. Constr. Build. Mater. **28**(1), 50–56 (2012)
14. T.T. Nguyen, Contribution à l'étude de la formulation et du procédé de fabrication d'éléments de construction en béton de chanvre, PhD thesis, (Bretagne-Sud University, France, 2010) p. 167
15. P. Tronet, T. Lecompte, V. Picandet, C. Baley, Study of lime hemp composite precasting by compaction of fresh mix—an instrumented die to measure friction and stress state. Powder Technol. **258**, 285–296 (2014)
16. P. Tronet, T. Lecompte, V. Picandet, C. Baylet, Study of lime and hemp concrete (lhc)—mix design, casting process and mechanical behaviors. Cem. Concr. Compos. **67**, 60–72 (2016)
17. S. Marceau, P. Glé, M. Guéguen-minerbe, E. Gourlay, S. Moscardelli, I. Nour, S. Amziane, Influence of accelerated aging on the properties of hemp concretes. Constr. Build. Mater. **139**, 524–530 (2016)

Chapter 2
Two Typical Plant Aggregates for Bio-Based Concretes

Hemp shiv (well-known) and rice husk (novel)

Many by-products of plant origin have been incorporated in mineral binders. However, it is important to distinguish plant fibers used as reinforcement in cement composite materials from plant-derived aggregates used for the manufacturing of lightweight insulating concretes (bio-based concretes). Those can be defined as the association of a high volume fraction of crop residues with a mineral binder [1]. This book does not deal with load-bearing concretes with very small amounts of aggregates or reinforcing fibers.

Most of the plant aggregates are derived from stems (hemp, flax, sunflower), straws (sorghum or miscanthus) and trunks (woodchips).

After grinding, the woody part of hemp stems gives rise to hemp shiv, a well-known aggregate associated with a lime-based binder to design Lime and Hemp Concrete (LHC). In order to diversify crop by-products, a novel kind of aggregate is explored. It corresponds to rice husk, the protective shell of rice grains. Since rice husk particles come from a totally different part of the plant, their characteristics will be presented and compared to those of hemp shives.

2.1 Source and Transformation Processes

2.1.1 Hemp Shiv

Hemp (*Cannabis Sativa*) is an annual plant whose height is from 1–3 m (Fig. 2.1a). This species is dedicated to the cultivation of industrial hemp in Central Asia and Europe. Hemp is grown as a break crop and harvested after 4 months of maturity [2, 3]. Thereafter, stems are cut and left on the field for a few weeks (retting). When the moisture content of stems is around 15%, those are harvested and taken to the defibering process to separate the woody part from the fibers (Fig. 2.1b).

© The Author(s) 2018
M. Chabannes et al., *Lime Hemp and Rice Husk-Based Concretes for Building Envelopes*, Biobased Polymers, https://doi.org/10.1007/978-3-319-67660-9_2

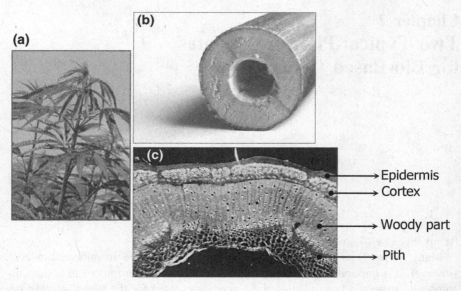

Fig. 2.1 **a** Hemp plant **b** Hemp stem **c** Micrograph of a cross-sectional view of a hemp stem colored with *green* Carmino of Mirande [3] (color figure online)

A micrograph of a cross-sectional view of a hemp stem colored with green Carmino of Mirande is reported in Fig. 2.1c. It makes it possible to describe cellulose-rich areas (in red) from those which are strongly lignified (in green) [3]:

- The epidermis consists of cells with cellulosic walls.
- The cortex contains the phloem fibers grouped into bundles.
- The woody part (from which the hemp shiv mainly comes) consists of parenchymal cells and xylem vessels.
- The cellulosic pith corresponds to the medullary parenchyma.

Plant tissues are rich in cellulosic compounds except the woody part for which the lignin content is clearly higher than in other areas (Fig. 2.1c). The woody part accounts for 50 wt% of the dry stem. As a result, the production yield of hemp shives is 3–3.5 T/ha (tons per hectare) or approximately 35,000 tons per year in France [4].

In France, the use of hemp shiv as lightweight aggregate for the manufacturing of LHC was initiated in the early 1990s.

After grinding, hemp shiv is in the form of chips whose morphological properties will be addressed. The shiv presented in this book is in compliance with the French professional rules for the construction of hemp concrete structures [5, 6] (Fig. 2.2).

The plant-based concretes studied by the authors will be manufactured with a shiv which either comes from FRD® (Troyes, France) or Technichanvre® (Riec-sur-Bélon, France).

Fig. 2.2 Commercial hemp
shiv

2.1.2   Rice Husk

According to figures from the International Grains Council (IGC) [7], rice is the
first cereal in the world for human food before wheat and corn. This is due to the
non-food use of rice which remains marginal if compared to wheat. Furthermore,
rice is locally grown in the South of France. It is particularly important to favor
local resources for the design of bio-based concretes as it is an opportunity to
develop eco-friendly materials by minimizing the impact of transport. In France,
rice fields are located in the Camargue area. The production of rough rice in France
is about 80,000 tons per year (according to a source in 2013) [8].

Rice harvest firstly consists in separating grains form straws and removing
impurities (insects, minerals and residues). Thereafter, rice grains undergo a drying
process before the husking where the outer covering is removed from the grain.
Rice husks are defined by two interlocking halves with a boat-like shape (lemma
and palea are botanical terms) for each rice grain [9]. When any other transformation process occurs, rice husk is qualified as natural. By contrast, some varieties
are parboiled. In the latter case, rice grains are soaked and exposed to water vapor
before the husking.

Rice husk (*Oryza Sativa*) can be considered as an agro-industrial by-product
coming from the rice hulling. This crop residue represents about 20 wt% of the
whole rough rice (i.e. paddy rice) harvested on the spikelets [10]. Hence, rice
farming produces nearly 15,000 tons of rice husks per year in France. Currently, the
use of rice husk is highly limited, this latter being regarded as a waste material often
buried in the ground or used as a fuel. Indeed, rice husk can be consumed for
electricity generation because of its high calorific value. However, the incineration
process is dangerous to human health and to the environment. Therefore, rice husk
causes critical problems in rice growing areas since significant volumes are generated and not used in a beneficial way [11]. In the building sector, the use of rice husk
ash as pozzolanic filler in cementitious binders was widely referenced [10, 12–16].

Fig. 2.3 Natural rice husk

Rice husk is characterized by a lower content of organic matter compared to other lignocellulosic by-products since it contains about 20% of amorphous silica [12, 17]. Consequently, when rice husk is burnt beyond 500 °C, the organic matter disappears and gives way to a SiO_2-rich nanometric ash with interesting pozzolanic properties [12, 16].

The use of whole rice husk as plant aggregate to design bio-based concretes is therefore totally novel. In this context, rice husks present many substantial advantages. They do not flame or smolder easily because of their particular silica-cellulose structural arrangement [18]. Moreover, given that husks do not biodegrade or burn easily, they are sometimes free-of-charge. Another asset is their availability throughout the year because most farms store rice and process it on a daily basis. Furthermore, it should be noted that hemp shives result from an industrial grinding process whereas rice husks can be used as they stand, requiring no shredding.

Natural rice husks presented here have not undergone any parboiling process and come from Biosud (Arles, France) (Fig. 2.3).

2.2 Chemical Composition

Lignocellulosic plants can be described at the cell wall scale. Cellulose and lignin represent about 70% of the plant biomass [1]. The structure of the cell wall is illustrated in Fig. 2.4 [19, 20].

The middle lamella (ML in Fig. 2.4a) is rich in pectin and acts as an adhesive between the cells. The primary cell wall (P) consists of cellulose microfibrils which lie parallel to each other. These chains of crystallized cellulose are embedded in an amorphous matrix of hemicelluloses (Fig. 2.4b). The secondary cell wall is divided

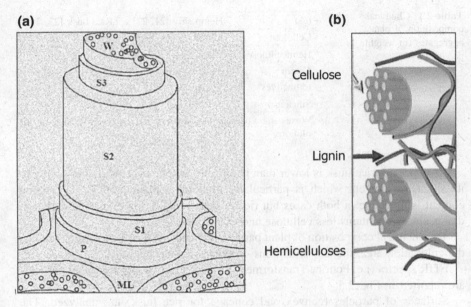

Fig. 2.4 a Cell wall of lignocellulosic plants **b** Arrangement of chemical components in the cell wall [20]

into three different layers (S_1, S_2 and S_3). It contains the same components than the primary cell wall but it is characterized by a higher lignin content [1, 3, 19, 20].

Plant-derived by-products have the same components (cellulose, hemicelluloses, lignin, pectins, waxes) but in variable proportions.

- Cellulose is a linear polysaccharide polymer with many glucose units. Owing to its strong inter and intra-bondings, mainly hydrogen bonds, cellulose is insoluble in most solvents. However, it is highly hydrophilic [1, 3, 19].
- Hemicelluloses are short-chained polysaccharides with an amorphous structure. Contrary to cellulose which contains only one sugar, hemicelluloses consist of several sugar units (glucose, galactose, mannose, arabinose, xylose, rhamnose) and uronic acids. Hemicelluloses are soluble in water and easily extracted from cell walls thanks to alkaline solutions. They are also hydrophilic [1, 3, 19].
- Lignin is a complex polymer with aliphatic and aromatic chains. The lignin content is especially high in the woody part of stems (xylem). Lignin provides stiffness and impermeability. It is effectively very hydrophobic (contrary to holocellulose) [1, 3].
- Pectins are acidic polysaccharides present in the middle lamella. Their carboxyl groups have a great capacity to exchange Ca^{2+} [3].
- Waxes consist of water-insoluble alcohols and acids. They can be extracted with organic solutions and are hydrophobic.

The chemical composition of plant aggregates coming from some studies is reported in Table 2.1.

Table 2.1 Chemical composition of plant aggregates (in weight percentages)

(%)	Hemp shiv [21, 22]	Rice husk [23, 24]
Cellulose	46–48	25–35
Hemicelluloses	12–21	18–21
Lignin	22–28	26–31
Extractives[a]	11–16	2–5
Silica ash	–	15–25

[a]Waxes, fats, pectins and sugars extracted by different aqueous solutions

As expected, rice husk is lower than hemp shiv in total organic matter owing to its silica ash content which is particularly high for a plant particle. The lignin content is the same in both cases but rice husk is actually lower in total carbohydrates since it contains less cellulose and extractives.

The chemical composition of plant particles depends on the plant variety but also the cultivation area, the soil conditions or even the plant maturity.

FTIR spectra (i.e. Fourier Transformed Infrared Spectroscopy) of plant particles are reported in Fig. 2.5.

Surfaces of particles (convex and concave for rice husk) are analyzed. The presence of amorphous polysaccharides such as hemicelluloses, pectins, waxes but also natural fats is highlighted by the presence of the band around 3300 cm^{-1} and peaks around 2900 and 1730 cm^{-1}. Lignin-associated absorbances are visible at 1600 and 1240 cm^{-1} as a peak for hemp shiv and a broad shoulder for rice husk. The peak at 1320 cm^{-1} corresponds to cellulose. For hemp shiv, the peak at

Fig. 2.5 FTIR spectra of plant particles. *P* polysaccharides, *WF* waxes and fats, *Pec* pectins, *H* hemicelluloses, *L* lignin, *C* cellulose, *Si* Silica [25]

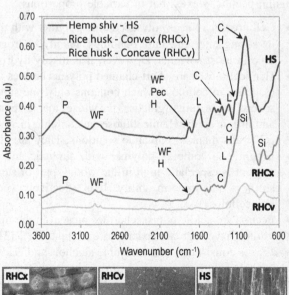

1030 cm^{-1} is associated to both hemicelluloses and cellulose. However, for rice husk, the broad band in the range $900–1200 \text{ cm}^{-1}$ is attributed to cellulosic compounds but also to silica. It was shown in Table 2.1 that rice husk contains less cellulose than hemp shiv but a significant quantity of silica on its surface. The peak corresponding to stretching vibrations of Si–C bonds at 800 cm^{-1} confirms the presence of silica for rice husk.

2.3 Physical and Morphological Properties

2.3.1 Microstructure

The cross-sectional view of a single particle by scanning electron microscopy (SEM) is reported in Fig. 2.6.

Figure 2.6b highlights the huge porosity of hemp shiv which is characterized by a honeycombed structure with lots of vessels whose diameter is more than 10 μm. The latter can be up to 60 μm [26]. The vessels are oriented longitudinally in the particle. The thickness of the walls between them is no more than 1 μm. The internal structure of a rice husk is different (Fig. 2.6a). Some little vessels can be observed but the solid phase appears to be in a majority proportion. Jauberthie et al. [12] and Park et al. [27] described the detailed microstructure of rice husk and the distribution of amorphous silica in the particle. Different SEM pictures are reported in Fig. 2.7.

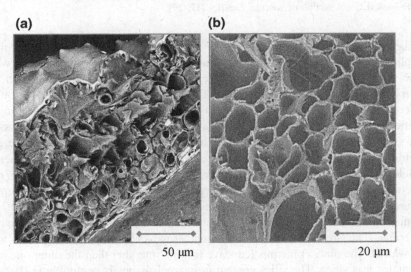

(a)   **(b)**

50 μm   20 μm

Fig. 2.6 SEM pictures of rice husk (**a**) and hemp shiv (**b**)

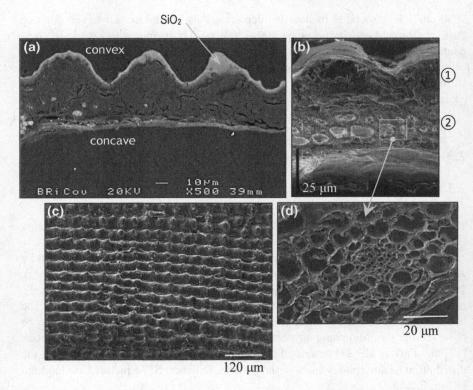

Fig. 2.7 **a** Silica distribution on a cross-sectional view of rice husk **b** Cross-sectional view of rice husk from the external epidermis to the internal epidermis **c** Morphology of the external epidermis of rice husk **d** Cross section of vascular bundles [12, 27]

The external surface (convex face) of the husk is characterized by a peculiar morphology. The epidermal cells are arranged in furrows and linear ridges with dome-shaped protrusions [27] (Fig. 2.7c). This external epidermis is rich in amorphous silica as shown in Fig. 2.7a. The layers of fibers underlying the outer epidermis are highly thick-walled according to Park et al. [27]. This area is strongly lignified and lowly porous (Fig. 2.7b, area n°1). The inner part of rice husk tissue shows a well-defined area with vascular bundles (Fig. 2.7b, area n°2). These are longitudinal vessels of phloem and xylem as they can be observed within the hemp particle. This area is significantly more porous than that consisted of the thick-walled epidermis and fibers. This is attested by Fig. 2.7d. From the external epidermis (convex) to the internal epidermis (concave), tissue organization consists of outer epidermis, layers of fibers, vascular bundles, parenchyma cells and inner epidermis. The area close to the inner epidermal cells is poorly lignified and thin-walled. The inner epidermis (concave face) is smoother than the outer one and it contains less silica. The silica content in internal tissues is negligible [12].

2.3.2 Densities and Porosities

The total porosity of bulk aggregates is related to the intergranular porosity and to the porosity within the particles. The knowledge of the bulk density, the apparent density of a particle (including the internal voids as presented in Fig. 2.8) and the true density of the solid phase make it possible to calculate the different porosities. These physical characteristics are all reported in Table 2.2.

Fig. 2.8 Schematic representation of a plant particle and its porosity

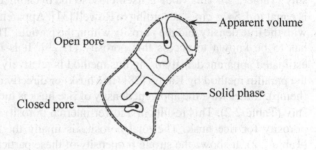

Open pore — Apparent volume

Solid phase

Closed pore —

Table 2.2 Densities and porosities of plant aggregates

Notation[a]	Method	Hemp shiv	Rice husk
		g cm^{-3}	
ρ_B	–	0.1	0.09
ρ_A	Estimated[b]	0.24	0.53
	Paraffin method [28]	–	0.65
	Stem section[c] [2]	0.26	–
ρ_T	Helium pycnometry[d]	–	1.48
	Air pycnometry [28]	–	1.42
	C_7H_8 pycnometry[e] [2]	1.47	–
		%	
η_O	$1 - \rho_A/\rho_T$	82–84	54–64
η_I	$1 - \rho_B/\rho_A$	58–62	83–86
η_T	$1 - \rho_B/\rho_T$	93	94

[a]Nomenclature
ρ_B—Bulk density
ρ_A—Apparent density (particle)
ρ_T—True density (solid phase)
η_O—Open porosity in the particle
η_I—Intergranular porosity
η_T—Total porosity in the bulk
[b]Particle density was calculated from the known pycnometric true density (ρ_T) and the water absorption capacity of particles after saturation (W_S) with this expression: $\eta_O = (W_S \times \rho_T)/[\rho_W + (W_S \times \rho_T)]$ where ρ_W is the density of water and W_S was taken as 120 wt% for rice husk [29] and 370 wt% for hemp shiv [1]
[c]Measurement on the basis of image analysis of a straight section of hemp stem
[d]Measured by the authors
[e]Toluene pycnometry

Table 2.2 shows that bulk density is about 0.1 g cm^{-3} for the two kinds of aggregates. Apparent and true densities are more difficult to measure. Some studies in literature report an underestimated value of the true density of rice husk [11, 29, 30] measured by the water displacement method. The latter has proved to be not accurate for plant particles as air is trapped in the pores [28]. The true density of rice husk is 1.48 g cm^{-3} according to Helium pycnometry. This result is very close to that reported in the work of Kaupp [28] in which air pycnometry has been used (1.42 g cm^{-3}). Moreover, the true density of rice husk is equivalent to that of hemp shiv (Table 2.2). This value is itself close to the pycnometric density of wood which is equal to 1.54 g cm^{-3} according to Rowell [31]. Apparent density can be calculated with the true density and the porosity within the particle. The rate of water absorption has to be known to access this porosity [1, 29]. It is interesting to see that the estimated apparent density using this method is relatively close to that measured by the paraffin method by Kaupp [28] (rice husk) or directly on the stem by Nguyen [2] (hemp). Either way, the apparent density of rice husk is more than twice that of hemp shiv (Table 2.2). This results in a lower internal porosity but a higher intergranular porosity for rice husk. The total porosity is finally the same for bulk aggregates (Table 2.2). It shows the strong propensity of these particles to provide a very good capacity for thermal insulation. It should be noted that the open porosity within hemp particles is very high and interconnected [32]. As regards rice husk, some authors [33, 34] refer to a certain amount of closed pores within the particle.

Based on the mercury porosimetry analysis, pore size distribution of particles is reported in Fig. 2.9. This method is only used to have a qualitative analysis of the pores within plant particles. It shows the three levels of porosity in hemp shiv. This is due to the structural organization of the xylem in which the conducting elements mainly consist of tracheid cells (from 5 to 50 μm) and vessels (from 50 to 300 μm). The punctuations which give the possibility for vessels to communicate can also create smaller pores. According to these results, rice husk is characterized by small pores under 0.1 μm. Almost no pores are detected from 1 to 30 μm. However, the presence of large vessels beyond 50 μm is detected. This is related to the region of vascular bundles which is common with hemp shiv.

Fig. 2.9 Pore size distribution by mercury porosimetry

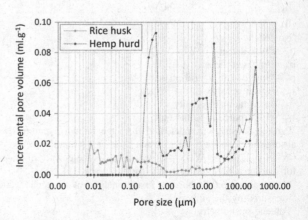

2.3.3 Shape and Particle Size Distribution

Particle size distribution of plant particles is performed by means of an image analysis processing. This method has proved its worth to obtain an efficient granulometric analysis of hemp aggregates [35] whereas some studies have shown the inconsistency of mechanical sieving for particulate materials with an elongated shape [36, 37]. Unlike the use of standard sieves which ultimately separate particles according to their width, image analysis allows to represent the minor and the major axis (width and length). It is well known that the cumulative passing curve obtained by mechanical sieving is actually identical to the width distribution plotted thanks to the results of image analysis [2]. In addition, the equivalent area diameter (EQ) can be calculated as described below [29] (Eq 2.1):

$$EQ = \sqrt{\frac{4 \times Area}{\pi}} \tag{2.1}$$

Figure 2.10 shows the size distribution of defibered hemp shives coming from two different suppliers. The length of the shives from Technichanvre® is marginally higher than that registered for the shiv of FRD®. The length distribution is about 2–25 mm in both cases.

A comparison between the size distribution of hemp shiv and rice husk is done in Fig. 2.11. Critical distinctions can be drawn from this analysis. The width of rice husk ranges from 1 to 4 mm and its maximum length is 10 mm. In the case of hemp shiv, the width distribution is about the same but the length can reach up to 25–27 mm (Fig. 2.11a). As a consequence, the elongation factor (EF), defined here

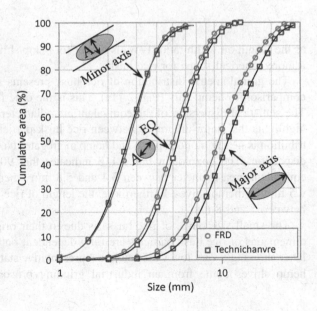

Fig. 2.10 Particle size distribution of hemp shives. Cumulative granulometric distribution by image analysis (ImageJ) for *minor axis* (width), equivalent area diameter and *major axis* (length). *A* Area. FRD and Technichanvre are two different French suppliers

Fig. 2.11 a Particle size
distribution of hemp shiv and
rice husk **b** Statistical
dispersion of the equivalent
area diameter (EQ), *RH* Rice
Husk,—*HS* Hemp Shiv

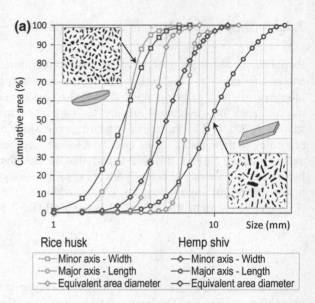

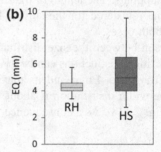

as the length on width ratio (L/W) is higher for hemp shiv (Table 2.3). This reflects
a more spherical shape for rice husk.

The granulometric distribution of rice husk presents a very small size range in
comparison to hemp shiv (Fig. 2.11a). This is the case for all the size parameters.
The statistical dispersion of the equivalent area diameter represented in Fig. 2.11b
highlights the major difference between rice husk particles for which the size distribution is nearly monodisperse and hemp aggregates characterized by a significant
dispersion. The boxplot in Fig. 2.11b indicates that 90% of rice husks have en
equivalent area diameter between 3.4 and 5.8 mm whereas it is between 2.8 and
9.5 mm for hemp shives. Furthermore, the length of rice husk is mainly distributed
between 5 and 8 mm.

The small size range of rice husks is due to their origin since it is known that
dimensions of rice husk particles are fully dependent on the size of rice grains, the
latter showing a very low variation in a representative statistical sample. In contrast,
hemp shives come from an industrial grinding process of hemp stems which

Table 2.3 Size parameters for a cumulative distribution of 50%. Median width (W_{50}), equivalent area diameter (EQ_{50}), length (L_{50}) and elongation factor (EF_{50})

Median parameters	Rice husk	Hemp shiv
W_{50} (mm)	2.8	2.7
EQ_{50} (mm)	4.3	5.2
L_{50} (mm)	6.7	10.3
EF_{50} (L_{50}/W_{50})	2.4	3.8

necessarily generates a more dispersed distribution. This more irregular shape is due to the shredding action of the size reduction machinery.

The particle size distribution provides valuable informations as it conditions the granular stacking of the plant-based concrete. Considering that the intergranular macroporosity of the concrete is linked to this granular stacking, the morphological parameters should have an influence on thermal and mechanical properties of plant-based concretes. Furthermore, a single husk is a flexible particle with a hollow semi-ellipsoidal shape and thickness lower than 100 μm (Fig. 2.6a). This geometry presents an original aspect when it is compared to the parallelepipedic shape of a hemp shiv particle (with a few millimeters thick) as regards the induced macroporosity in the concrete.

2.4 Specific Properties for Plant-Based Concretes

2.4.1 Water Absorption Capacity

Figure 2.12 reports the water absorption capacity of plant aggregates. The test consisted in immersing dried particles (48 h at 105 °C) in water during 8 h and

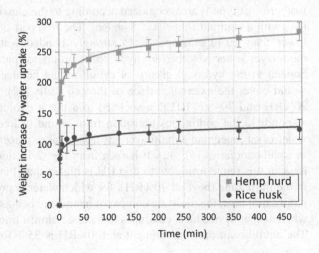

Fig. 2.12 Mass relative water absorption of plant aggregates

following the mass uptake registered at appropriate intervals of time. Particles were put in a spherical strainer immersed in water. For each measurement, the latter was introduced into a spinner and centrifuged one minute.

For hemp shiv like rice husk, absorption kinetics is very fast during the first few minutes. The absorption rate of hemp shiv after 5 min is 200% according to Chabannes et al. [29] (Fig. 2.12). It is between 225 and 280% for other authors with different shives and various methods to test the water absorption [1, 2, 29, 38, 39]. The absorption rate of rice husk after 5 min is 100%. Beyond 5 min, a second stage takes place where the water uptake strongly decreases. During this stage, rice husks are quickly close to saturation when hemp shives keep absorbing a higher amount of water. It has been shown that hemp shiv is able to absorb nearly 4 times its dry weight after 48 h of immersion before saturation (370% [1]). Figure 2.12 shows that weight gain of immersed rice husk is about 120% after 8 h. According to Tamba et al. [40], the water absorption rate of rice husk after 20 h of immersion before saturation is 80%. Another author reports a rate of 122% [18]. No details are provided on the husk variety (e.g. parboiled or not).

One of the difficulties encountered for designing plant-based concretes is the competition for water between the mineral binder and plant aggregates during the manufacture of the material. A lack of water for the binder due to water absorption by plant vessels can disrupt the setting of the concrete.

2.4.2   Sorption Isotherms

The use of plant aggregates with high porosity brings to consider their behavior towards water vapor. It is well known that moisture content of porous materials increases with ambient relative humidity, especially that crop by-products are hydrophilic.

Sorption isotherms of plant aggregates at 25 °C are reported in Fig. 2.13. Isotherms of type II are recognized according to the classification of Brunauer et al. [41] with a sigmoid shape (i.e. S-shape). The first part of the profiles (here for RH lower than 20%) corresponds to strongly bonded water (structural water) and monolayer water adsorbed by hydrophilic groups of particles (water is hydrogen-bonded to the hydroxyl groups of cellulose and hemicelluloses) [41]. This water would cover the external surface of the cell walls [42]. The second part between 20%RH and 70–80%RH is associated to a linear evolution of the profile. It corresponds to the continuous transition from bound to free water. The first layer of water is saturated and a multilayer adsorption takes place. The water is then present in small capillaries [41, 42]. It is seen from Fig. 2.13 that moisture content of rice husk in this range from low to mid RH is higher than that of hemp shiv. However, in the third zone (beyond 70%RH), for which water is present in the liquid state in large capillaries, the moisture content of hemp shiv begins to increase more strongly whereas that of rice husk continues to increase almost linearly up to final saturation. The equilibrium moisture content at 100%RH is 35% for hemp shiv when that of

Fig. 2.13 Sorption isotherms of rice husk [44] and hemp shiv [39]

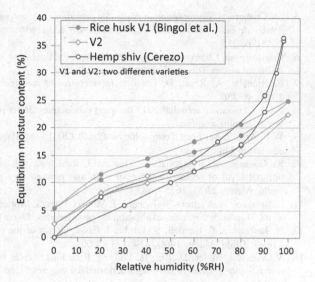

rice husk is lower than 25%. These differences between the two kinds of particle are related to their porous structure but also to their chemical composition. Xylem vessels of hemp shiv are thick-walled tubes consisting of a polysaccharide-rich primary wall surrounding a lignified secondary cell wall. Hemp shiv is therefore highly hydrophilic and composed of large vessels as demonstrated before. This probably explains the higher sorption of water vapor in hemp shives for high RH. By contrast, the outer layer of rice husk is covered with waxes, protective pectin, silica and lignin. As a result, it is quite hydrophobic if it is compared to hemp shiv [43]. The behavior of rice husk for low RH could be explained by the presence of extremely small pores in outer layers of rice husk.

Water vapor sorption of plant aggregates has a role to play in the drying process of plant-based concretes but also towards carbonation and internal curing [45].

References

1. V. Nozahic, Vers une nouvelle démarche de conception des bétons végétaux lignocellulosiques basée sur la compréhension et l'amélioration de l'interface Liant/Végétal. Application à des granulats de chènevotte et de tige de tournesol associés à un liant ponce/chaux, Ph.D. Thesis (Clermont University, France, 2012), p. 311
2. T.T. Nguyen, Contribution à l'étude de la formulation et du procédé de fabrication d'éléments de construction en béton de chanvre, Ph.D. thesis (Bretagne-Sud University, France, 2010) p. 167
3. D. Sedan, Etude des interactions physico-chimiques aux interfaces fibres de chanvre/ciment. Influence sur les propriétés mécaniques du composite, Groupe d'étude des Matériaux Hétérogènes, Ph.D. Thesis, (Limoges University, France, 2007) p. 129
4. CETIOM, Enquête culturale, Chanvre 2013. Available: http://www.cetiom.fr

5. M. Chabannes, *Formulation et étude des propriétés mécaniques d'agrobétons légers isolants à base de balles de riz et de chènevotte pour l'éco-construction* (Université de Montpellier, 2015), p. 215
6. Constuire en Chanvre, *Règles professionnelles d'éxécution* (SEBTP. Société d'Édition du Bâtiment et des Travaux Publics, 2012)
7. E. Biénabe, A. Rival, D. Loeillet, *Développement durable et filières tropicales* (QUAE, 2016), p. 336
8. FAO, Classement mondial 2013 des pays producteurs de riz paddy, (2014) Available: http://www.lasyntheseonline.fr
9. W.P. Armstrong, Fruit Terminology—Part 2, (2001). Available: http://waynesword.palomar.edu/termfr2.htm
10. K. Ganesan, K. Rajagopal, K. Thangavel, Rice husk ash blended cement: assessment of optimal level of replacement for strength and permeability properties of concrete. Constr. Build. Mater. **22**(8), 1675–1683 (2008)
11. T. Serrano, M. Victoria Borrachero, J. Monzó, J. Payà, Morteros aligerados con cascarilla de arroz: diseño de mezsclas evaluación de propriedades. Dyna **175**, 128–136 (2012)
12. R. Jauberthie, F. Rendell, S. Tamba, I. Cisse, Origin of the pozzolanic effect of rice husks. Constr. Build. Mater. **14**(8), 419–423 (2000)
13. V. Van, C. Rößler, D. Bui, H. Ludwig, Rice husk ash as both pozzolanic admixture and internal curing agent in ultra-high performance concrete. Cem. Concr. Compos. **53**, 270–278 (2014)
14. P. Chindaprasirt, S. Homwuttiwong, C. Jaturapitakkul, Strength and water permeability of concrete containing palm oil fuel ash and rice husk-bark ash. Constr. Build. Mater. **21**(7), 1492–1499 (2007)
15. G.C. Cordeiro, R.D. Toledo Filho, L.M. Tavares, E.D.M.R. Fairbairn, S. Hempel, Influence of particle size and specific surface area on the pozzolanic activity of residual rice husk ash. Cem. Concr. Compos. **33**(5), 529–534 (2011)
16. W. Xu, Y.T. Lo, D. Ouyang, S.A. Memon, F. Xing, W. Wang, X. Yuan, Effect of rice husk ash fineness on porosity and hydration reaction of blended cement paste. Constr. Build. Mater. **89**, 90–101 (2015)
17. N. Johar, I. Ahmad, A. Dufresne, Extraction, preparation and characterization of cellulose fibres and nanocrystals from rice husk. Ind. Crops Prod. **37**(1), 93–99 (2012)
18. M. González De la Cotera, Morteros Ligeros de Cáscara de Arroz, in *IV Congreso Nacional de Ingeniería Civil*, 1982
19. M. Ibrahim Nasr Morsi, Properties of rice straw cementitious composite, PhD Thesis (Darmstadt University, Germany, 2011) p. 147
20. D.M. Alonso, S.G. Wettstein, J.A. Dumesic, Bimetallic catalysts for upgrading of biomass to fuels and chemicals. Chem. Soc. Rev. **41**(24), 8075–8098 (2012)
21. Y. Diquélou, E. Gourlay, L. Arnaud, B. Kurek, Impact of hemp shiv on cement setting and hardening: influence of the extracted components from the aggregates and study of the interfaces with the inorganic matrix. Cem. Concr. Compos. **55**, 112–121 (2014)
22. C. Garcia-Jaldon, D. Dupeyre, M.R. Vignon, Fibres from semi-retted hemp bundles by steam explosion treatment. Biomass Bioenerg. **14**(3), 251–260 (1998)
23. K.G. Mansaray, A.E. Ghaly, Thermal degradation of rice husks in an oxygen atmosphere. Energy Sources **21**(5), 453–466 (1999)
24. T.P.T. Tran, J.-C. Bénézet, A. Bergeret, Rice and Einkorn wheat husks reinforced poly(lactic acid) (PLA) biocomposites: effects of alkaline and silane surface treatments of husks. Ind. Crops Prod. **58**, 111–124 (2014)
25. M. Chabannes, E. Garcia-Diaz, L. Clerc, J.C. Bénézet, Effect of curing conditions and Ca (OH)2-treated aggregates on mechanical properties of rice husk and hemp concretes using a lime-based binder. Constr. Build. Mater. **102**, 821–833 (2016)
26. P. Glé, Acoustique des Matériaux du Bâtiment à base de Fibres et Particules Végétales. Outils de Caractérisation, Modélisation et Optimisation, PhD Thesis (École Nationale des Travaux Publics de l'État, France, 2013) p. 127

27. B.-D. Park, S.G. Wi, K.H. Lee, A.P. Singh, T.-H. Yoon, Y.S. Kim, Characterization of anatomical features and silica distribution in rice husk using microscopic and micro-analytical techniques. Biomass Bioenerg. **25**(3), 319–327 (2003)
28. A. Kaupp, *Gasification of rice hulls: theory and Praxis* (Vieweg + Teubner Verlag, Wiesbaden, 1984)
29. M. Chabannes, J.-C. Bénézet, L. Clerc, E. Garcia-Diaz, Use of raw rice husk as natural aggregate in a lightweight insulating concrete: an innovative application. Constr. Build. Mater. **70**, 428–438 (2014)
30. J. Salas and J. Veras Castro, Materiales de construcción con propiedades aislantes a base de cascara de arroz. Inf. la Constr. **37**(372), 53–64 (2012)
31. R. Rowell, Moisture properties, in *Handbook of wood chemistry and wood composites* (CRC Press, Boca Raton, 2005), p. 21
32. F. Collet, M. Bart, L. Serres, J. Miriel, Porous structure and water vapour sorption of hemp-based materials. Constr. Build. Mater. **22**(6), 1271–1280 (2008)
33. A. Prada, C.E. Cortés, La descomposición térmica de la cascarilla de arroz: una alternativa de aprovechamiento integral. Rev. ORINOQUIA **14**(1), 155–170 (2010)
34. A. Kaupp, J. Goss, Technical and economical problems in the gasification of rice hulls. Physical and chemical properties. Energy Agric. **1**, 201–234 (1983)
35. V. Nozahic, S. Amziane, G. Torrent, K. Saïdi, H. De Baynast, Design of green concrete made of plant-derived aggregates and a pumice–lime binder. Cem. Concr. Compos. **34**(2), 231–241 (2012)
36. V. Picandet, P. Tronet, C. Baley, *Caractérisation granulométrique des chènevottes, in 30e Rencontres AUGC-IBPSA* (Chambéry, France, 2012)
37. C. Igathinathane, L.O. Pordesimo, E.P. Columbus, W.D. Batchelor, S. Sokhansanj, Sieveless particle size distribution analysis of particulate materials through computer vision. Comput. Electron. Agric. **66**(2), 147–158 (2009)
38. L. Arnaud, E. Gourlay, Experimental study of parameters influencing mechanical properties of hemp concretes. Constr. Build. Mater. **28**(1), 50–56 (2012)
39. V. Cerezo, Propriétés mécaniques, thermiques et acoustiques d'un matériau à base de particules végétales : approche expérimentale et modélisation théorique, PhD Thesis (École Nationale des Travaux Publics de l'État, Lyon, France 2005), p. 243
40. S. Tamba, I. Cisse, F. Rendell, R. Jauberthie, *Rice husk in lightweight mortars, in Second international symposium on structural lightweight aggregate concrete* (Kristiansand, Norway, 2000), pp. 117–124
41. R.D. Andrade, R. Lemus, C. Pérez, Models of sorption isotherms for food. Vitae **18**, 325–334 (2011)
42. M.V. Bastias, A. Cloutier, Evaluation of wood sorption models for high temperatures. Maderas Ciencias y Tecnol. **7**(3), 145–158 (2005)
43. A. Bazargan, T. Gebreegziabher, C.-W. Hui, G. McKay, The effect of alkali treatment on rice husk moisture content and drying kinetics. Biomass Bioenerg. **70**, 468–475 (2014)
44. G. Bingol, B. Prakash, Z. Pan, Dynamic vapor sorption isotherms of medium grain rice varieties. LWT—Food Sci. Technol. **48**(2), 156–163 (2012)
45. P. Lura, M. Wyrzykowski, C. Tang, E. Lehmann, Internal curing with lightweight aggregate produced from biomass-derived waste. Cem. Concr. Res. **59**, 24–33 (2014)

Chapter 3
Lime-Based Binders

Unlike Portland cement, the use of lime goes back much further in time. During the Greco-Roman period, walls were built of lime-based mortars blended with fine sand and pozzolanic additives (volcanic ash). Initially, only pure limestone was extracted from quarries to produce calcic lime. It was not until the beginning of the 19th century that Vicat introduced the hydraulicity index of building limes which were then classified according to the clay content of limestone. Thereafter, hydraulic lime, derived from the calcination of argillaceous limestone, was widely used prior to the development of Ordinary Portland Cement (OPC) [1].

Nowadays, in spite of the hegemony of OPC, building limes are experiencing a revival in the repair and restoration of old buildings but also for eco-construction.

3.1 Production and General Properties

3.1.1 Calcic Lime

Calcic lime (also known as hydrated lime, aerial lime or air lime) comes from pure limestone or including potentially less than 5% of magnesium oxide (MgO).

It is produced in two stages. The first one is the decarbonation of calcium carbonate ($CaCO_3$) at 900 °C (3.1):

$$CaCO_3 \rightarrow CaO + CO_2 \qquad (3.1)$$

Then, quicklime (CaO) is slaked with water. This operation leads to the production of hydrated lime as calcium hydroxide $Ca(OH)_2$ according to the following Eq. (3.2):

© The Author(s) 2018
M. Chabannes et al., *Lime Hemp and Rice Husk-Based Concretes for Building Envelopes*, Biobased Polymers, https://doi.org/10.1007/978-3-319-67660-9_3

$$CaO + H_2O \rightarrow Ca(OH)_2 \tag{3.2}$$

When the slaking process is controlled, a lime powder is obtained. Different kinds of Calcic Limes (CL) are defined depending on CaO and MgO contents. The most widely used is CL90–S which contains at least 80% of $Ca(OH)_2$ in the final product (Table 3.1) [2].

Physical properties of CL90–S are reported in Table 3.2. Aerial lime is a binder with a low bulk density and a high specific surface area. The latter can be up to about 15 times higher than that of OPC [3]. According to Pavia et al. [4], most lime particles are sized between 10 and 100 μm. Some authors [5, 6] agree that aerial lime is mostly composed of micrometric particles of size in the range of 1–20 μm that could be ascribed to portlandite crystals. The median diameter (D_{50}) has been assessed to be 10 μm by Cardoso et al. [6].

3.1.2 Hydraulic Lime

Hydraulic lime is obtained from calcareous-siliceous rocks. In this case, limestone is partly composed of reactive silicates. During the burning process at approximately 1200 °C, calcium oxide (CaO) is combined with silica to produce calcium silicates. Decarbonation of $CaCO_3$ and slaking of CaO are made in the same way as for calcic lime. The final binder is a lime including both $Ca(OH)_2$ and C_2S (dicalcium silicates) also known as belite (Fig. 3.1) [1, 7].

Chemical composition of the raw material (CaO/SiO_2 ratio) will influence the proportion of C_2S in the lime binder and consequently its hydraulic properties. Natural Hydraulic Limes are designated as NHL. They are assigned a specific number (NHL2, NHL3.5, NHL5) corresponding to their minimum compressive strength at 28 days in relation to their hydraulic index. For instance, NHL3.5 is a moderately hydraulic lime with an intermediate C_2S content. Its compressive strength after 28 days is above 3.5 MPa according to building lime standard EN 459. Chemical and mineralogical composition of hydraulic limes is reported in Table 3.3. As the strength class of NHL increases, the C_2S content increases and the free $Ca(OH)_2$ decreases. Some aluminate components (Al_2O_3) can be present in tiny amounts (1–2%) [8, 9].

Physical properties of NHL are reported in Table 3.4. Densities are higher than that of CL90–S. According to Arizzi et al. [5], specific surface area of hydraulic limes is intermediate between that of OPC and that of calcic lime.

Table 3.1 Chemical and mineralogical composition of calcic lime powder CL90-S in weight percentages [2, 20, 48]

Chemical		Mineralogical	
CaO	LOI[a]	Ca(OH)$_2$	Unburnt CaCO$_3$
65–75	25–27	80–90	5–10

[a]Loss on ignition

Table 3.2 Physical properties of CL90-S

Bulk density (kg m^{-3}) [3, 62]	400–450
Absolute density (kg m^{-3}) [62, 63]	2200–2500
BET—SSA (m^2 g^{-1})^a [3–6, 64]	14–18
Grain size distribution (μm) [5]	0.1–100

^{a}SSA Specific Surface Area

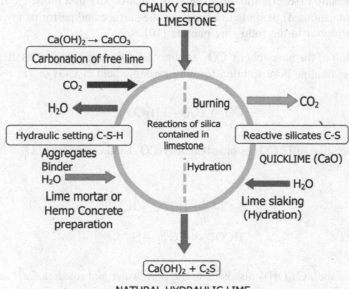

Fig. 3.1 Cycle of hydraulic lime [7]

Table 3.3 Chemical and mineralogical composition of Natural Hydraulic Limes (NHL) in weight percentages [5, 8, 9, 21, 48, 65]

Chemical			Mineralogical		
CaO	SiO$_2$	LOI^a	Ca(OH)$_2$	C$_2$S	CaCO$_3$
50–70	6–20	15–20	30–50	20–40	5–20

^{a}Loss on ignition

Table 3.4 Physical properties of NHL

Bulk density (kg m^{-3}) [8, 9]	500–850
Absolute density (kg m^{-3}) [63]	2500–2700
BET—SSA (m^2 g^{-1}) [5]	9.26
Grain size distribution (μm) [5, 66]	0.1–100

3.2 Hardening Mechanisms

3.2.1 Aerial Carbonation

Aerial lime-based mortars set by evaporation of excess free water and carbonation due to the presence of atmospheric carbon dioxide. Carbonation is a very slow acid-base reaction (several months even years) that occurs in a moist environment (wet but not saturated). It starts to happen on the surface and performs towards the core of the mortar in the following manner [10]:

(i) Diffusion of the atmospheric CO_2 gas into the pore structure and its dissolution into the alkaline pore solution forming carbonic acid H_2CO_3 (3.3):

$$CO_2 + H_2O \rightarrow H_2CO_3 \tag{3.3}$$

(ii) Dissociation of H_2CO_3 as bicarbonate (HCO_3^-) and carbonate (CO_3^{2-}) ions (3.4 and 3.5):

$$H_2CO_3 \leftrightarrow HCO_3^- + H^+ \tag{3.4}$$

$$HCO_3^- \leftrightarrow CO_3^{2-} + H^+ \tag{3.5}$$

(iii) In parallel, $Ca(OH)_2$ dissolves in the pore water and releases Ca^{2+} and OH^- ions (3.6):

$$Ca(OH)_2 \leftrightarrow Ca^{2+} + 2OH^- \tag{3.6}$$

(iv) Reaction between Ca^{2+} and CO_3^{2-} ions leading to the precipitation of $CaCO_3$ crystals through nucleation and crystal growth (3.7):

$$Ca^{2+} + (OH^-)_2 + 2H^+ + CO_3^{2-} \leftrightarrow CaCO_3 + 2H_2O \tag{3.7}$$

The overall reaction is the following (3.8):

$$Ca(OH)_2 + CO_2 \leftrightarrow CaCO_3 + H_2O + 74\,kJ/mol \tag{3.8}$$

Nucleation and growth of $CaCO_3$ in the smallest pores leads to the decrease in the microporosity with time and subsequent increase in the bulk density of the lime-based mortar.

A number of factors control the carbonation mechanisms and kinetics. They are linked to the diffusion process of CO_2 (1) and chemical reaction between the dissolved CO_2 and $Ca(OH)_2$ (2).

Fig. 3.2 Diffusion of CO_2 gas in the pore system of lime mortars [10]

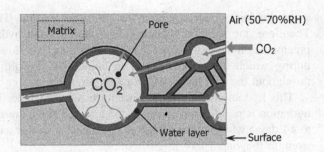

1. Diffusion of gaseous CO_2 depends on the pore system (amount, size, distribution), degree of saturation of pore spaces, curing conditions (relative humidity and CO_2 concentration in the ambient air), initial water-to-binder mass ratio, drying rate or even binder content [10–14]. Water-saturated pores hinder CO_2 diffusion and carbonation reaction cannot proceed.
2. Chemical reactivity of CO_2 is influenced by binder type (mineral phases) and concentration. The high $Ca(OH)_2$ content of calcic limes contributes to promote the carbonation process [12]. Moreover, the high specific surface area (SSA) of $Ca(OH)_2$ particles is also a factor for the CO_2 reactivity [10]. The amount of water in pores has further to be considered as the presence of water in the pore system is a prerequisite for the dissolution of reactants.

To sum up, carbonation is only sustained if a free path exists for CO_2 to move into the lime mortar and if water is present at the same time for $Ca(OH)_2$ and CO_2 to dissolve. A water-layer in pore walls is necessary (Fig. 3.2). For porous materials, pore water is directly linked to relative humidity (RH) in the ambient air. When relative humidity is very high (RH > 90%), capillary condensation results in a blockage against CO_2 diffusion in the interconnected porous medium. Carbonation is possible for a relative humidity between 40 and 90%RH [10]. Cizer et al. [10] explain that such conditions help to preserve a water layer in pores with a thickness between 0.4 and 0.8 nm. However, it is recognized that the relative humidity range for optimal carbonation is 50–70%RH [11, 12].

3.2.2 Hydraulic Setting Due to C₂S Hydration

Hydration of C_2S in hydraulic limes is performed in the following way (3.9) [15]:

$$2C_2S + 4H_2O \rightarrow C_3-S_2-H_3 + CH^* \qquad (3.9)$$

*C: CaO, S: SiO₂, H: H₂O according to the cement chemist notation

Hydration of calcium silicate grains is performed through a dissolution/precipitation process. When mixing water is added, grains are dissolved and there is a growing ionic concentration in the interstitial medium ($H_2SiO_4{}^{2-}$, Ca^{2+} and OH^-). Therefore, the solution becomes supersaturated and hydration products (C–S–H) precipitate on the grains surface [16]. The setting is followed by a hardening period during which hydration kinetics is limited by the diffusion of water and ions throughout the first layers of hydration products [15].

This hydration reaction also occurs in OPC together with C_3S hydration. C_3S hydration is rapid and contributes to early age strength of cement (from few hours to 14 days) whereas that of C_2S is significantly slower and contributes to later age strength (beyond 7 days).

Goñi et al. [17] conducted a quantitative analysis of the hydration of C_3S and C_2S pastes cured at $\sim 100\%RH$. The mass fraction of all the mineral components of hardened pastes are reported in Fig. 3.3.

These results show the significant amount of unreacted C_2S after 28 days compared to C_3S. It should also be noted that the amount of portlandite formed (CH) is very low in the case of C_2S paste (less than 4% after 28 days). Authors have also determined the hydration degree of the pastes. Hydration kinetics of C_2S is considerably slower than that of C_3S. Nevertheless, this difference tends to diminish at later stages of hydration. The hydration degree of C_2S paste after 1 year appears to be no more than 0.8 (Fig. 3.4) [17].

There is a paucity of studies on C_2S hydration in NHL binders. Xu et al. [18] showed that the water-to-binder ratio has a strong impact on the hydration rate of hydraulic lime. They also pointed out that hydration of NHL proceeds gradually from 1 h up to 3 days of curing but their study is not carried out over a longer period. However, Lanas et al. [19] stated that the major part of C_2S contribution to the strength of NHL-based mortars occurs beyond 28 days. In the two previous studies, lime mortars were cured at about 60–70%RH and 20–25 °C.

Fig. 3.3 Mass fraction distribution of all the components of C_2S (**a**) and C_3S (**b**) pastes determined from quantitative TGA [17]

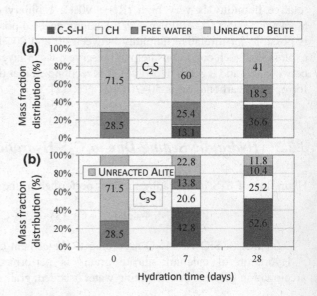

Fig. 3.4 Hydration degree of pastes with hydration time [17]

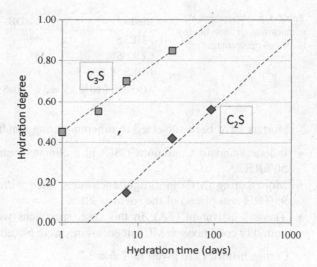

In light of previous developments, hardening process of hydraulic limes by C_2S hydration and aerial carbonation is slow and proceeds up to 1 year even more. However, these hardening mechanisms are strongly impacted by curing conditions such as temperature, relative humidity and CO_2 concentration.

3.3 Influence of Curing Conditions on Hardening

3.3.1 Effect of Relative Humidity and Temperature

Hardening of NHL–based mortars depending on curing conditions has been studied by a few authors [20–22]. Mineralogical properties of the binder after hardening have been investigated by X-ray diffraction (XRD), thermogravimetric analysis (TGA) but also back-scattered scanning electron microscopy (BSE–SEM). Grilo et al. [21] and Arizzi et al. [22] stored NHL3.5 mortars at 60–65%RH and compared their mineralogical composition after hardening with that of mortars cured under humid conditions (90–95%RH). They found that C_2S hydration is particularly favored in the case of high RH. Chabannes et al. [20] studied the effect of both relative humidity and curing temperature on the hardening of various lime mortars for which the composition of unhydrated binders is reported in Table 3.5.

Mix proportions of lime mortars prepared in accordance with EN 459 standard are reported in Table 3.6. All mortars were mixed with the same aggregate-to-lime mass ratio of 2.75. The amount of water was set in order to achieve a mortar flow of 165 ± 10 mm after the flow table test (EN 459-2).

Table 3.5 Mineralogical composition of lime binders (in weight percentages)

Binder	Ca(OH)$_2$	C$_2$S	CaCO$_3$
NHL3.5	45	~30	8–10
CL90–S	85	–	8–10
NHL3.5/CL90–S[a]	65	~15	8–10

[a]Mixture of NHL3.5 and CL90–S at 50/50 wt%

Mortars have been subjected to different curing conditions:

- Indoor Standard Conditions (ISC) in a climate-controlled room at 20 °C and 50%RH.
- Moist curing (MC) in airtight enclosures filled with 5 cm of water to ensure 95%RH and placed in the room at 20 °C.
- Thermal activation (TA). In this case, specimens were cured under the same humidity conditions as MC but enclosures were placed in the oven at 50 ± 2 °C.

Curing histories are given in Table 3.7.

TGA curves of powdered matrix samples collected in the bulk of NHL3.5 mortars cured under different environments until 28 days are reported in Fig. 3.5. Water bound to C–S–H can be determined by performing analysis of these curves between 100 and 400 °C [23]. Results indicate that water bound to C–S–H is higher for the mortars subjected to moist curing (7d–MC and 21d–MC) than those cured under indoor standard conditions (28d–ISC). Nevertheless, water bound to C–S–H for mortars cured at 50 °C cannot be compared with that of mortars cured at 20 °C since the water content of C–S–H cured at high temperature is known to be lower. According to many authors [24–26], when calcium silicate binders are cured under elevated temperature, C–S–H phases are crystallized in a different manner with a modified Ca/Si ratio. They also exhibit a high decrease in their water content and increase in their density.

In addition, it should be noted that the same trend is obtained for NHL3.5/CL90–S mortars (which are less rich in C$_2$S).

Amounts of Ca(OH)$_2$ and CaCO$_3$ for the mortars including C$_2$S phases are given in Table 3.8. Carbonation rates are also deduced. 28d–ISC samples are more carbonated than all other samples, which is not surprising since pore saturation with water during moist curing hinders carbonation and drying kinetics of mortars at 50%RH is fast. However, the amount of Ca(OH)$_2$ for 21d–MC and 7d–TA samples

Table 3.6 Mix proportions of lime-based mortars

Mortar type	Mass ratios		Composition (g)		
	A/L[a]	W/B[b]	Lime	Aggregates	Water
NHL3.5	2.75	0.56	392	1080	220
CL90–S	2.75	0.71	392	1080	278
NHL3.5/CL90–S	2.75	0.62	392	1080	243

[a]A/L Aggregate-to-lime
[b]W/B Water-to-binder

Table 3.7 Curing histories of lime mortars (*d* days, *ISC* Indoor Standard Conditions, *MC* Moist curing, *TA* Thermal activation)

Age	1d	7d	21d	28d
28d–ISC	20 °C–50%RH			
7d–MC	20 °C–95%RH	20 °C–50%RH		
7d–TA	50 °C–95%RH	20 °C–50%RH		
21d–MC	20 °C–95%RH		20 °C–50% RH	

Fig. 3.5 TGA curves of powdered matrix samples collected in the bulk of NHL3.5 mortars after 28 days

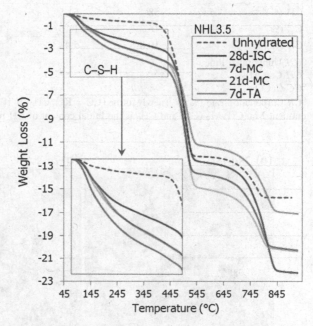

is very low compared to the amount of $CaCO_3$ (Fig. 3.5, Table 3.8). In view of the binder deposit in enclosures, dissolution of the binder in wet conditions is assumed. Binder leaching in calcium-based materials is a well-known phenomenon for mortars exposed to high humidity. Due to the high porosity of lime mortars, condensed water (RH > 95%) may penetrate into the material, causing leaching of Ca $(OH)_2$. Then, soluble components can migrate through the material to be deposited as efflorescence on the external surfaces of the mortar and in the enclosure [27–29]. Forster et al. have well demonstrated that $Ca(OH)_2$ is effectively vulnerable to dissolution in uncarbonated NHL mortars [27]. With the same $CaCO_3$ content than mortars cured under high RH and room temperature, the amount of $Ca(OH)_2$ in mortars cured under high temperature is further reduced. The influence of temperature on the leaching of $Ca(OH)_2$ is linked to its solubility product but also to the diffusivity of calcium ions Ca^{2+} throughout the lime mortar. Even if solubility of Ca $(OH)_2$ is known to decline as temperature increases, the increase in temperature can accelerate leaching due to a rising diffusivity of Ca^{2+} [27].

BSE-SEM images of NHL3.5 mortars in polished cross sections are represented in Fig. 3.6 for 28d–ISC and 7d–TA samples.

Table 3.8 Results from TGA performed in the bulk of mortars including C_2S

Curing	UH^a	28d–ISC	7d–MC	21d–MC	7d–TA
Mortar	NHL3.5				
CH (%)^b	45	37	43.6	37	28.8
CC (%)^c	8.5	19.3	10.5	12	11.4
CR (%)^d	–	17.8	3.1	5.8	4.8
Mortar	NHL3.5/CL90–S				
CH (%)^a	65	58	61.5	56	51.5
CC (%)^b	8.5	16	11.5	13	14
CR (%)^c	–	10.8	5.4	5.1	6.3

^{a}*UH* Unhydrated lime

^{b}*CH* $Ca(OH)_2$

^{c}*CC* $CaCO_3$

^{d}*CR* Carbonation rate (%) = $[(\text{newly formed CC} \times K_3)/CH_0] \times 100$ where K_3 is the molar mass quotient $M_W(CH)/M_W(CC)$ and CH_0 is the initial content of CH present in unhydrated lime

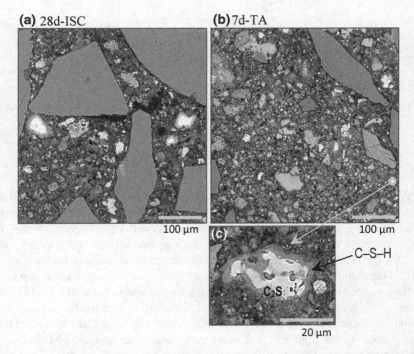

Fig. 3.6 BSE-SEM images of polished cross sections of NHL3.5 mortars after 28 days. **a** 28d–ISC mortar, **b** 7d–TA mortar, **c** C_2S grain with C–S–H rim in 7d–TA mortar

A number of unhydrated cores of C_2S (white in BSE) surrounded by C–S–H rims (light grey) can be observed for 28d–ISC mortars (Fig. 3.6a). However, it becomes harder to find unhydrated C_2S for 7d–TA mortars (Fig. 3.6b). It means that remaining unhydrated C_2S are smaller due to the thermo-activation of hydration reactions. This leads to thicker C–S–H rims. Furthermore, 7d–TA samples seem to be characterized by a denser microstructure.

Some unhydrated cores with their hydration rim can be observed with a higher magnification (Fig. 3.6c).

The preceding results show that C_2S hydration can be accelerated in lime-based mortars even with a rather low amount of C_2S ($\sim 15\%$) corresponding to a mortar mixed with NHL3.5 and CL90–S at 50/50 wt%. In the latter case and obviously for an increasing proportion of C_2S, moist curing (>95%RH) and elevated temperature (>20 °C) are curing conditions that promote hardening due to a faster formation of C–S–H gel.

This being said, the hardening process by carbonation is optimal in the range 50–70%RH as previously stated. The works of Cizer et al. [30] have shown that the hardening of binders including calcium hydroxide and calcium silicates can suffer from a competition effect between carbonation and hydration.

3.3.2 Effect of CO_2 Concentration

Natural carbonation is a slow process as a consequence of the low concentration of CO_2 in the air (0.03–0.04%) [11]. The exposure of lime-based mortars to higher CO_2 concentrations could be effective to accelerate the strength development of these mortars at early age (after a few weeks). The accelerated carbonation test is often used to study the durability of cementitious materials according to XP P18-458 standard. A conservation of the specimens at 20 °C and 65%RH in a curing enclosure with 50% of CO_2 is recommended according to Turcry et al. [31]. There are a number of investigations about natural carbonation of lime-based mortars, especially those of Lawrence et al. [32–34]. They study the carbonation depth of mortars by means of the phenolphthalein test (Fig. 3.7) and present also an accurate method to evaluate the local carbonation rate in mortars using both phenolphthalein and TGA. However, little research has been done about accelerated carbonation of lime-based mortars. Only the study of Cultrone et al. [35] was interested in comparing mortars cured under natural conditions with those cured

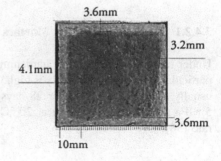

Fig. 3.7 Phenolphthalein test on the cross section of a lime-based mortar (colorless area is carbonated and *pink* area is uncarbonated or partially carbonated [33]) (color figure online)

under accelerated carbonation. The carbonation rate is investigated with TGA, XRD but also with the weight gain of samples. According to these authors [35], a 90 wt% $Ca(OH)_2$–$CaCO_3$ transformation can be achieved in over one week following the curing of mortars in a specific enclosure at 25 °C–50%RH and CO_2-saturated. Thereafter, a sharp fall in the CO_2 consumption occurred. A high concentration of CO_2 is suspected to release an excessive heat (since carbonation is exothermic as seen in 3.8) involving a premature drying of samples and the disruption of disso-lution mechanisms of CO_2 [35]. The release of water during the carbonation process and the enclosing of $Ca(OH)_2$ particles by an impervious shell of $CaCO_3$ are other relevant factors which could prevent complete carbonation [33].

3.4 Physico-Mechanical Properties After Hardening

3.4.1 Porosity and Hygrothermal Properties

In a calcic lime paste (CL90–S), the capillary porosity is high and strongly inter-connected. According to Lanas and Alvarez [36], open porosity of mortars incor-porating CL90–S binder ranges from 25 to 50%. Pore size distribution of calcic lime pastes is characterized by a pore diameter ranging from 0.5 to 1 μm depending on the water-to-binder ratio [37]. It also includes pores of very small size (~ 10 nm) related to the porosity of crystals.

The dry thermal conductivity of CL90–S mortars is between 0.67 and 0.84 W m^{-1} K^{-1} for a bulk density around 1720 kg m^{-3} [38, 39]. Furthermore, aerial limes are known to provide high water vapor permeability. It should be noted that permeability of lime mortars is 10 times higher than that of cement mortars [15].

Porosity of hydraulic lime mortars is more or less identical to that of calcic lime mortars but the average pore diameter is lower in the case of hydraulic lime [40]. The thermal conductivity of NHL-based mortars remains close to that of CL90–S mortars [15]. By contrast, hydraulic lime mortars are less permeable to water vapor according to Silva et al. [40].

3.4.2 Mechanical Properties

3.4.2.1 Aerial Lime-Based Mortars and Pozzolanic Binders

Under natural conditions, the hardening of aerial lime-based mortars is very slow, particularly if relative humidity is high. The compressive strength of such mortars is usually less than 2 MPa after 28 days of hardening [20]. However, it can reach 5 MPa after 1 year of curing at 20 °C–60%RH due to the advanced carbonation rate [36].

It is possible to improve the hardened properties of aerial lime-based mortars with adequate proportions of pozzolanic additives. These are natural pozzolans (pumice), calcined pozzolans (such as metakaolin) or industrial by-products (fly ash and silica fume) [15]. Many studies deal with aerial lime-based mortars blended with metakaolin [41–46]. Metakaolin comes from the burning of milled kaolinite at 650–800 °C. It can be considered as an amorphous inorganic material and is composed of 54% in mass of SiO_2 and 46% in mass of Al_2O_3 (AS2) [44]. The reaction of metakaolin with $Ca(OH)_2$ forms secondary C–S–H (II) which provide mechanical strength in the short and medium term. This kind of pozzolanic reaction is written as follows (3.10) [15]:

$$AS_2 + 5CH + 5H \rightarrow C_5AS_2H_{10} \qquad (3.10)$$

where $C_5AS_2H_{10}$ is in reality an average composition of the hydration products. When the lime is in excess, two newly-formed phases corresponding to C–S–H and C_4AH_{13} are present with $Ca(OH)_2$. By contrast, if the lime is fully consumed, C_2ASH_8 will be also present.

The strength gain is linked to the pozzolanic effect but also to the increased compactness of the mix provided by metakaolin [15]. Lime-metakaolin mortars are sensitive to curing conditions. Their hardening under moist conditions (high RH) is significantly faster than that of pure aerial lime-based mortars. According to many authors [42, 43, 46] the addition of 20 wt% of metakaolin (in relation to the mass of aerial lime) is found to be optimal to increase the compressive strength of aerial lime-based pastes and mortars. Velosa et al. [44] highlight the increase in compressive strength of lime mortars (volumic ratio CL90–S/MK = 2) when they incorporate metakaolin. Results depend on the type of metakaolin used. Its chemical composition has a strong influence on its reactivity and consequently on the strength development (Fig. 3.8).

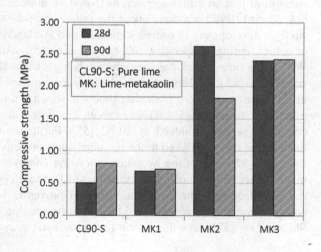

Fig. 3.8 Compressive strength of lime-metakaolin mortars (MK1, 2 and 3 for different metakaolins depending on their mineralogical composition) compared to that of pure lime mortars (CL90–S) [44]

Formulated lime is frequently used for the manufacturing of hemp concretes. The commercial binder called Tradical® PF70 consists of hydrated lime with hydraulic binder and pozzolanic material. It falls under the group FLA3.5 according to EN 459 standard dealing with building limes. It means that compressive strength of standardized mortar using this binder should be at least equal to 3.5 MPa after 28 days [2]. Nguyen et al. [47] report a compressive strength of almost 10 MPa for the paste with a water-to-lime mass ratio of 0.5. According to the producer, this binder contains 80% of hydrated lime by volume, the remainder being attributed to hydraulic binders. The mineralogical composition has been investigated in the Ph. D. of Dinh [48]. $Ca(OH)_2$ (35 wt%), C_2S but also C_3S have been found as main phases. The specific density of this binder is 2450 kg m^{-3} [49, 50]. It is close to that of feebly hydraulic limes (NHL2).

It has been seen that C_2S hydration in hydraulic lime mortars can be accelerated when the curing temperature is 50 °C. This is also the case for the hydration of C_3S in Portland cement. The sensitivity of the hardening kinetics to the curing temperature is described by the Arrhenius law which introduces the apparent activation energy. The latter represents the dependence of the temperature to reactions and mechanisms which run during the hydration process of the binder [51]. The reaction rate f(T) is expressed as follows [52] (3.11):

$$f(T) = A \times \exp\left(\frac{-E_A}{RT}\right) \qquad (3.11)$$

where A is a proportionality constant, E_A is the apparent activation energy (kJ mol^{-1}), R is the gas constant (J K^{-1} mol^{-1}) and T is the hydration temperature (Kelvin).

The apparent activation energy will depend upon the mineralogical composition and the grinding size of the hydraulic binder.

High curing temperature is frequently used to increase the early age compressive strength of precast building materials (based on mineral aggregates). Many studies [24, 53–56] have been done about the influence of the curing temperature on the strength development of ordinary concrete and Portland cement mortars. The effect of a high curing temperature (40–60 °C) is clearly visible until about 3–7 days. Within this time period, the strength gain is higher than that observed for materials cured at 20 °C [55]. Nevertheless, the high temperature is responsible for changes in the microstructure of hydration products which are more heterogeneously distributed in the matrix [56]. This can result in a lower ultimate compressive strength compared to that achieved at 20 °C [55]. Furthermore, the effect of the heat treatment is strongly linked to the so-called apparent activation energy of the binder used. It is known that the hydration activation energy of lime-pozzolan blends is much higher (66 kJ mol^{-1} [57]) than that of Portland cements (40 kJ mol^{-1} [52]). As a consequence, the pozzolanic reaction proves to be more sensitive to high curing temperature if compared to the hydration reaction of C_3S [57]. The works of Shi and Day [57] show that curing lime-pozzolan pastes (20% of aerial lime, 80%

Fig. 3.9 Effect of curing
temperature on the
compressive strength
development of
lime-pozzolan pastes
(RH > 95%) [57]

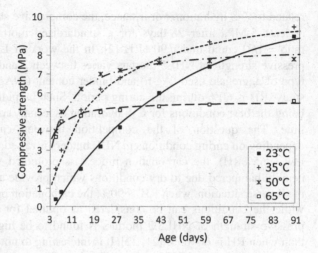

of natural pozzolan) between 35 °C and 65 °C provide a strength gain which is higher than that registered at 20 °C at least up to 30 days (Fig. 3.9).

The slower strength development of the pozzolanic binder compared to that of an ordinary Portland cement allows observing the effect of temperature on hydration kinetics in the longer term. Pastes cured at 23 °C only reach 4 MPa after 28 days while a compressive strength of 7 MPa is achieved for those cured at 35 °C (Fig. 3.9). However, the strength development of lime pastes cured at high temperature (i.e., 35 °C and more) tends to level off much more quickly than those cured at 23 °C. For the curing at 65 °C, the ultimate compressive strength is achieved for about 5 MPa. By contrast, the compressive strength continues to increase beyond 8 MPa after 90 days of curing for the paste cured at the ambient temperature (Fig. 3.9).

Another attractive possibility to enhance the compressive strength of lime-based materials at early age is to accelerate the carbonation process. This has been investigated by the use of chemical additives in some studies of Medici et al. [58, 59]. An amine-based resin reacting with acid gas like CO_2 has been successful in accelerate the carbonation of aerial lime-based mortars and pastes, thus enhancing their short-term compressive strength. Accelerated carbonation using a CO_2-rich atmosphere remains more widespread. This aspect has been addressed by Cultrone et al. [35] from a microstructural point of view. The effect on the compressive strength of lime-based pastes has not been reported.

3.4.2.2   Hydraulic Lime Mortars

Mechanical properties of hydraulic limes depend on the hydraulic index, the water-to-lime mass ratio and the curing conditions. These parameters are directly related to C_2S hydration. For instance, the compressive strength of NHL5 should be at least 5 MPa after 28 days for a standardized mortar (binder/aggregate mass ratio = 1/3) cured under 90%RH [2]. In the work of Lanas et al. [19], the compressive strength of NHL5 mortars varies between 2 and 9 MPa depending on the type of aggregate used (Ag-n) and binder content (B/Ag) (Fig. 3.10). In the latter study, RH is 60% during the curing period. Such humidity conditions are far from being the best conditions for C_2S hydration especially for this eminently hydraulic lime. The question of the competition between carbonation and hydration depending on curing conditions in NHL binders is raised. When relative humidity is low (<65%RH), the carbonation process is promoted whereas C_2S hydration is heavily hampered due to dry conditions which provide a harmful lack of water. In the reverse situation, when RH > 90%, the carbonation process is strongly hindered while the conditions can be considered as optimal for C_2S hydration. The compressive strength of NHL3.5 mortars is found to be higher when RH is 90–95% than when RH is 60–65% [21, 22]. It is interesting to note that EN 459 standard [2] recommends to cure NHL2 under optimal conditions for carbonation (i.e., 60–65% RH) while a RH higher than 90–95% is recommended for NHL3.5 and NHL5.

Figure 3.11 reports the compressive strength of lime-based mortars (CL90–S, NHL3.5 and a blend of both limes) cured 28 days under different environments [20]. Mineralogical composition (Table 3.5), mix proportions of mortars (Table 3.6) and details regarding curing conditions (Table 3.7) have been mentioned above.

For lime-based mortars including hydraulic phases (NHL3.5 and NHL3.5/CL90– S), the lowest compressive strength is achieved for samples cured 28 days under indoor standard conditions just after the demolding (28d–ISC). By contrast, initial moist curing (7d–MC) leads to a higher compressive strength. For NHL3.5 mortars, the latter is 2.6 times higher than that reached after 28 days under ISC. The increase is more significant for the extended moist curing (21d–MC) for which compressive strength is 7.5 MPa for NHL3.5 mortars. Results are in accordance with EN 459 standard indicating that compressive strength of NHL3.5 mortars after 28 days ranges

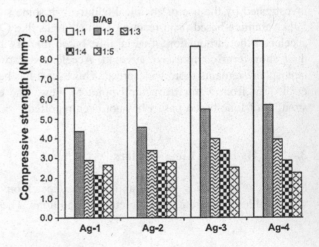

Fig. 3.10 Compressive strength of NHL5 mortars after 28 days of curing at 60% RH as a function of the aggregate type and binder-to-aggregate (B/Ag) mass ratio [19]

Fig. 3.11 Compressive strength of lime-based mortars cured under different environments

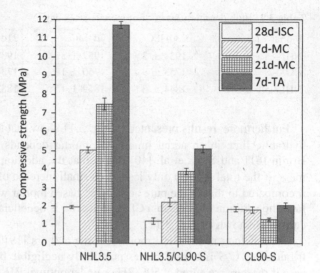

from 3.5 to 10 MPa for mortars prepared with a binder-to-sand mass ratio of 3 and cured at 20 °C and over 90%RH (moist curing) [2]. Moreover, it is confirmed that high RH promotes the hardening by the hydration of C_2S whereas dry conditions (50%RH) slow down hydration reactions due to a strong lack of water. In addition, Fig. 3.11 particularly stresses the effect of the initial curing (i.e., the first 7 days) on the strength development of lime-based mortars including C_2S. Samples cured under moist conditions and 50 °C during 7 days (7d–TA) exhibit the highest compressive strength after 28 days. The effect of elevated temperature is highlighted by comparison of 7d–TA with 7d–MC samples. Compressive strength of activated mortars (7d–TA) is about 2.3 times higher than that reached under 7d–MC. It increased from 5 to 11.7 MPa for NHL3.5 mortars and from 2.2 to 5.1 for NHL3.5/CL90–S mortars. This is attributed to the thermo-activation of C_2S hydration. It should be noted that even with a low fraction of C_2S, compressive strength of NHL3.5/CL90–S mortars remain very sensitive to RH and temperature.

As regards calcic lime mortars (CL90–S), compressive strength is almost not impacted by curing regimes (Fig. 3.11). Results show that curing CL90–S mortars during 7 days under moist conditions does not change compressive strength after 28 days compared to the standard curing. Only extended moist curing (21d–MC) leads to a slight decrease in strength certainly attributed to a lower carbonation rate. Pore saturation during 21 days gives priority to the transfers in liquid phase which are much slower than those in gas phase [60].

Bulk densities of mortars after 28 days are reported in Table 3.9.

An increase in bulk densities of lime mortars including hydraulic phases can be noted for moist curing at 20 °C. This is attributed to the higher water consumption for C_2S hydration. For mortars cured under elevated temperature, the bulk density cannot be compared with that of specimens cured under 20 °C as C–S–H water content is known to be different.

Table 3.9 Bulk densities (kg m^{-3}) of mortars at 28 days

Mortar	28d–ISC	7d–MC	21d–MC	7d–TA
NHL3.5	1934 ± 3	1957 ± 5	1988 ± 2	1989 ± 6
CL90–S	1753 ± 9	1750 ± 12	1738 ± 22	1746 ± 8
NHL3.5/CL90–S	1844 ± 3	1859 ± 15	1882 ± 14	1860 ± 2

Furthermore, results presented in Fig. 3.11 show that the effect of adding natural hydraulic lime in an aerial lime-based mortar depends on curing conditions. De Bruijn [61] and Silva et al. [40] report that the addition of NHL5 (up to 75% by mass of the total binder) only leads to a small increase of the compressive strength if compared to that of the pure aerial lime-based mortar with a curing at 60%RH. In fact, the addition of NHL in a CL90–S mortar is beneficial only in the case of moist curing as shown in Fig. 3.11.

To conclude about NHL mortars, some authors [19] explain that strength contribution of C_2S is considered as practically negligible before 28 days. This is the case if mortars are cured at 50%RH or under optimal RH for carbonation (60–65%). However, high RH (>95%) and elevated temperature strongly accelerate C_2S hydration from the first days, resulting in the increase of compressive strength at early age (quite dramatically at 95%RH and 50 °C).

References

1. G. Martinet, P. Souchu, "La chaux - Définitions et histoire," *Techniques de l'Ingénieur* (2009), pp. 11, C922
2. NF EN 459-1 Standard. Building Lime—Part 1, AFNOR (2015)
3. M. Fourmentin, P. Faure, S. Gauffinet, U. Peter, D. Lesueur, D. Daviller, G. Ovarlez, P. Coussot, Porous structure and mechanical strength of cement-lime pastes during setting. Cem. Concr. Res. **77**, 1–8 (2015)
4. S. Pavia, R. Walker, P. Veale, A. Wood, Impact of the properties and reactivity of rice husk ash on lime mortar properties. J. Mater. Civ. Eng. **26**, (2014)
5. A. Arizzi, G. Cultrone, M. Brummer, H. Viles, A chemical, morphological and mineralogical study on the interaction between hemp hurds and aerial and natural hydraulic lime particles: Implications for mortar manufacturing. Constr. Build. Mater. **75**, 375–384 (2015)
6. F.A. Cardoso, H.C. Fernandes, R.G. Pileggi, M.A. Cincotto, V.M. John, Carbide lime and industrial hydrated lime characterization. Powder Technol. **195**(2), 143–149 (2009)
7. CESA, St Astier Pure and NHLs. Available: http://www.stastier.co.uk/nhl/info/hydraul.htm
8. J. Grilo, P. Faria, R. Veiga, A. Santos Silva, V. Silva, A. Velosa, New natural hydraulic lime mortars—Physical and microstructural properties in different curing conditions. Constr. Build. Mater. **54**, 378–384 (2014)
9. P.F.G. Banfill, A.M. Forster, S. Mackenzie, M.P. Sanz, E.M. Szadurski, Natural hydraulic limes for masonry repair : Hydration and workability. 34th Cem. Concr. Sci. Conf. **224**, (2014)
10. Ö. Cizer, K. Van Balen, J. Elsen, D. Van Gemert, Real-time investigation of reaction rate and mineral phase modifications of lime carbonation. Constr. Build. Mater. **35**, 741–751 (2012)

11. S. Bernal, J. Provis, R. Mejia de Gutierrez, J. Van Deventer, Accelerated carbonation testing of alkali-activated slag/metakaolin blended concretes: Effect of exposure conditions. Mater. Struct. **48**, 653–669 (2015)
12. M. Fernández Bertos, S.J.R. Simons, C.D. Hills, P.J. Carey, A review of accelerated carbonation technology in the treatment of cement-based materials and sequestration of CO_2. J. Hazard. Mater. **112**(3), 193–205 (2004)
13. C. Shi, F. He, Y. Wu, Effect of pre-conditioning on CO_2 curing of lightweight concrete blocks mixtures. Constr. Build. Mater. **26**(1), 257–267 (2012)
14. A. Morandeau, M. Thiéry, P. Dangla, Investigation of the carbonation mechanism of CH and C–S–H in terms of kinetics, microstructure changes and moisture properties. Cem. Concr. Res. **56**, 153–170 (2014)
15. S. Amziane, L. Arnaud, *Les bétons de granulats d'origine végétale* (Lavoisier. France, Application au béton de chanvre, 2013)
16. A. Gmira, Étude structurale et thermodynamique d'hydrates modèle du ciment, Ph.D. Thesis, Orleans University, France, p. 215, 2004
17. S. Goñi, F. Puertas, M.S. Hernández, M. Palacios, A. Guerrero, J.S. Dolado, B. Zanga, F. Baroni, Quantitative study of hydration of C3S and C2S by thermal analysis. J. Therm. Anal. Calorim. **102**(3), 965–973 (2010)
18. W. Xu, Y.T. Lo, D. Ouyang, S.A. Memon, F. Xing, W. Wang, X. Yuan, Effect of rice husk ash fineness on porosity and hydration reaction of blended cement paste. Constr. Build. Mater. **89**, 90–101 (2015)
19. J. Lanas, J.L.P. Bernal, M. Bello, J.I. Galindo, Mechanical properties of natural hydraulic lime-based mortars. Cem. Concr. Res. **34**, 2191–2201 (2004)
20. M. Chabannes, E. Garcia-Diaz, L. Clerc, J.C. Bénézet, Effect of curing conditions and Ca $(OH)_2$-treated aggregates on mechanical properties of rice husk and hemp concretes using a lime-based binder. Constr. Build. Mater. **102**, 821–833 (2016)
21. J. Grilo, A. Santos Silva, P. Faria, A. Gameiro, R. Veiga, A. Velosa, Mechanical and mineralogical properties of natural hydraulic lime-metakaolin mortars in different curing conditions. Constr. Build. Mater. **51**, 287–294 (2014)
22. A. Arizzi, G. Martinez-Huerga, E. Sebastián-Pardo, G. Cultrone, Mineralogical, textural and physical-mechanical study of hydraulic lime mortars cured under different moisture conditions. Mater. Construcción **65**(318), e053 (2015)
23. S. Xu, J. Wang, Y. Sun, Effect of water binder ratio on the early hydration of natural hydraulic lime. Mater. Struct. **48**(10), 3431–3441 (2014)
24. C. Famy, K.L. Scrivener, A. Atkinson, A.R. Brough, Effects of an early or a late heat treatment on the microstructure and composition of inner C–S–H products of Portland cement mortars. Cem. Concr. Res. **32**(2), 269–278 (2002)
25. A. Gajewicz, *Characterisation of Cement Microstructure and Pore—Water Interaction by 1H Nuclear Magnetic Resonance Relaxometry* (University of Surrey, UK, 2014), p. 162
26. J. Escalante-Garcia, J. Sharp, Variation in the composition of C–S–H gel in portland cement pastes cured at various temperatures. J. Am. Ceram. Soc. **82**(11), 3237–3241 (1999)
27. A.M. Forster, E.M. Szadurski, P.F.G. Banfill, Deterioration of natural hydraulic lime mortars, I: Effects of chemically accelerated leaching on physical and mechanical properties of uncarbonated materials. Constr. Build. Mater. **72**, 199–207 (2014)
28. T. Ekström, *Leaching of Concrete. Experiments and Modelling* (Division of Building Materials, LTH, Lund University, 2001), p. 200
29. A. Cheng, S.-J. Chao, W.-T. Lin, Effects of leaching behavior of calcium ions on compression and durability of cement-based materials with mineral admixtures. Mater. (Basel) **6**(5), 1851–1872 (2013)
30. O. Cizer, Competition between carbonation and hydration on the hardening of calcium hydroxide and calcium silicate binders, Ph.D. Thesis, Catholic University of Leuven, Belgium, 2009
31. P. Turcry, L. Oksri-Nelfia, A. Younsi, A. Aït-Mokhtar, Analysis of an accelerated carbonation test with severe preconditioning. Cem. Concr. Res. **57**, 70–78 (2014)

32. R.M.H. Lawrence, A study of carbonation in non-hydraulic lime mortars, Ph.D. Thesis, University of Bath, UK, p. 316, 2006
33. R.M.H. Lawrence, T.J. Mays, P. Walker, D. D'Ayala, Determination of carbonation profiles in non-hydraulic lime mortars using thermogravimetric analysis. Thermochim. Acta **444**(2), 179–189 (2006)
34. R.M. Lawrence, T.J. Mays, S.P. Rigby, P. Walker, D. D'Ayala, Effects of carbonation on the pore structure of non-hydraulic lime mortars. Cem. Concr. Res. **37**(7), 1059–1069 (2007)
35. G. Cultrone, E. Sebastián, M.O. Huertas, Forced and natural carbonation of lime-based mortars with and without additives: Mineralogical and textural changes. Cem. Concr. Res. **35** (12), 2278–2289 (2005)
36. J. Lanas, J.I. Alvarez, Masonry repair lime-based mortars: Factors affecting the mechanical behavior. Cem. Concr. Res. **33**(11), 1867–1876 (2003)
37. M. Arandigoyen, J.L.P. Bernal, M.A.B. López, J.I. Alvarez, Lime-pastes with different kneading water: Pore structure and capillary porosity. Appl. Surf. Sci. **252**(5), 1449–1459 (2005)
38. R. Cerny, Z. Pavlik, M. Pavlikova, Hygric and thermal properties of materials of historical masonry, in *Proceedings on the 8th symposium on building physics in the Nordic Countries*, Technical University of Denmark, 2008, pp. 903–910
39. E. Vejmelková, M. Keppert, P. Rovnaníková, Z. Keršner, R. Černý, Application of burnt clay shale as pozzolan addition to lime mortar. Cem. Concr. Compos. **34**(4), 486–492 (2012)
40. B. Silva, A.P. Ferreira Pinto, A. Gomes, Influence of natural hydraulic lime content on the properties of aerial lime-based mortars. Constr. Build. Mater. **72**, 208–218 (2014)
41. M.R. Veiga, F. Carvalho, Some performances characteristics of lime mortars for use on rendering and repointing of ancient buildings, in *5th International Masonry Conference*, London, 1998, pp. 107–111
42. E. Vejmelková, R. Pernicová, R. Sovják, R. Černý, Properties of innovative renders on a lime basis for the renovation of historical buildings, in *Structural studies, repairs and maintenance of heritage architecture XI*, 2009, pp. 221–229
43. E. Vejmelková, M. Keppert, Z. Keršner, P. Rovnaníková, R. Černý, Mechanical, fracture-mechanical, hydric, thermal, and durability properties of lime-metakaolin plasters for renovation of historical buildings. Constr. Build. Mater. **31**, 22–28 (2012)
44. A. Velosa, F. Rocha, R. Veiga, Influence of chemical and mineralogical composition of metakaolin on mortar characteristics. Acta Geodyn. e Geomater. **6**(1), 121–126 (2009)
45. M. Stefanidou, Study of the microstructure and the mechanical properties of traditional repair mortars, Ph.D. Thesis, Department of Civil Engineering, University of Thessaloniki, Greece, 2000
46. A. Arizzi, G. Cultrone, Aerial lime-based mortars blended with a pozzolanic additive and different admixtures: A mineralogical, textural and physical-mechanical study. Constr. Build. Mater. **31**, 135–143 (2012)
47. T.-T. Nguyen, V. Picandet, S. Amziane, C. Baley, Influence of compactness and hemp hurd characteristics on the mechanical properties of lime and hemp concrete. Eur. J. Environ. Civ. Eng. **13**(9), 1039–1050 (2009)
48. T.M. Dinh, Contribution au développement du béton de chanvre préfabriqué utilisant un liant pouzzolanique innovant, Ph.D. Thesis, Toulouse 3 University (Paul Sabatier), France, p. 211, 2014
49. P. Tronet, T. Lecompte, V. Picandet, C. Baley, Study of lime hemp composite precasting by compaction of fresh mix—An instrumented die to measure friction and stress state. Powder Technol. **258**, 285–296 (2014)
50. P. Tronet, T. Lecompte, V. Picandet, C. Baylet, Study of lime and hemp concrete (lhc)—Mix design, casting process and mechanical behaviors. Cem. Concr. Compos. **67**, 60–72 (2016)
51. S. Siddiqui, *Effect of Temperature and Curing on the Early Hydration of Cementitious Materials* (Bangladesh University of Engineering and Technology, 2010). p. 169

52. A. Kouakou, C. Legrand, E. Wirquin, Mesure de l'énergie d'activation apparente des ciments dans les mortiers à l'aide du calorimètre semi-adiabatique de Langavant. Mater. Struct. **29**, 444–447 (1996)
53. K.O. Kjellsen, R.J. Detwiler, O.E. Gjørv, Pore structure of plain cement pastes hydrated at different temperatures. Cem. Concr. Res. **20**(6), 927–933 (1990)
54. J.-K. Kim, Y.-H. Moon, S.-H. Eo, Compressive strength development of concrete with different curing time and temperature. Cem. Concr. Res. **28**(12), 1761–1773 (1998)
55. T. Boubekeur, K. Ezziane, E.-H. Kadri, Estimation of mortars compressive strength at different curing temperature by the maturity method. Constr. Build. Mater. **71**, 299–307 (2014)
56. E. Gallucci, X.Y. Zhang, K. Scrivener, Influence de la température sur le développement microstructural des bétons, *Septième édition des journées scientifiques du regroupement francophone pour la recherche et la formation sur le béton (RF)²B* (France, Toulouse, 2006), p. 10
57. C. Shi, R.L. Day, Acceleration of strength gain of lime-pozzolan cements by thermal activation. Cem. Concr. Res. **23**, 824–832 (1993)
58. F. Medici, L. Piga, G. Rinaldi, Behaviour of polyaminophenolic additives in the granulation of lime and fly-ash. Waste Manag. **20**(7), 491–498 (2000)
59. F. Medici, G. Rinaldi, Poly-Amino-Phenolic additives accelerating the carbonation of hydrated lime in mortar. Environ. Eng. Sci. **19**(4), 271–276 (2002)
60. L. Arnaud, E. Gourlay, Experimental study of parameters influencing mechanical properties of hemp concretes. Constr. Build. Mater. **28**(1), 50–56 (2012)
61. P. De Bruijn, *Hemp Concrete: Mechanical Properties Using Both Shives and Fibers* (Faculty of Landscape planning. Swedish University of Agricultural Sciences, Lund, 2008)
62. V. Nozahic, S. Amziane, G. Torrent, K. Saïdi, H. De Baynast, Design of green concrete made of plant-derived aggregates and a pumice–lime binder. Cem. Concr. Compos. **34**(2), 231–241 (2012)
63. J. Chamoin, Optimisation des propriétés (physiques, mécaniques et hydriques) de bétons de chanvre par la maîtrise de la formulation, Ph.D. Thesis, Rennes 1 University, INSA Rennes, France, p. 198, 2013
64. Y. Sébaïbi, R.M. Dheilly, B. Beaudoin, M. Quéneudec, The effect of various slaked limes on the microstructure of a lime-cement-sand mortar. Cem. Concr. Res. **36**(5), 971–978 (2006)
65. M. Chabannes, E. Garcia-Diaz, L. Clerc, J.-C. Bénézet, Studying the hardening and mechanical performances of rice husk and hemp-based building materials cured under natural and accelerated carbonation. Constr. Build. Mater. **94**, 105–115 (2015)
66. A. Bras, F.M.A. Henriques, Natural hydraulic lime based grouts—The selection of grout injection parameters for masonry consolidation. Constr. Build. Mater. **26**(1), 135–144 (2012)

Chapter 4
Lime and Hemp or Rice Husk Concretes for the Building Envelope: Applications and General Properties

4.1 Applications in Buildings, Casting Processes and Mix Design of Plant-Based Concretes

4.1.1 Application of Hemp Concrete in Housing

Hemp concrete is obtained from the mix of hemp shiv, water and a mineral binder (which can be itself a mixture of different binders). Professional rules for the construction of hemp concrete structures [1] report the main characteristics and the implementation of the material depending on the specific application. Hemp concrete is used for walls (filling of outer walls, doubling of load-bearing walls, etc.), roof insulation or even floor slabs (Fig. 4.1) [2].

Design bulk density, thermal conductivity and minimum required mechanical performances of the main applications are presented in Table 4.1. The binder content of hemp concrete used for roof insulation is very low (from about 100 to 120 kg m^{-3}). This kind of mix has only an hygrothermal function since the mechanical strength is highly limited. The mixture intended to be used as a filling material in a wall timber frame (i.e., WALL mixture) is a good compromise between thermal insulation and mechanical strength. It is defined by an intermediate shiv content (from 120 to 150 kg m^{-3}) and a binder content of about 300 kg m^{-3} [2, 3, 4].

The following developments will focus more specifically on lime and hemp concrete (LHC) for wall applications. LHC are typically used as filling materials manually tamped in timber stud walls (cast-in-place in the form of successive beds) (Fig. 4.2a). Precast blocks can also be manufactured by static loading or vibro-compaction of the freshly-mixed material (Fig. 4.2b). Actually, this method opens up interesting ways of improving the compressive strength of LHC. Lastly, for the rehabilitation of old buildings, sprayed hemp concrete is another option (Fig. 4.2c).

© The Author(s) 2018
M. Chabannes et al., *Lime Hemp and Rice Husk-Based Concretes for Building Envelopes*, Biobased Polymers, https://doi.org/10.1007/978-3-319-67660-9_4

Fig. 4.1 Applications of hemp concrete in housing [2]

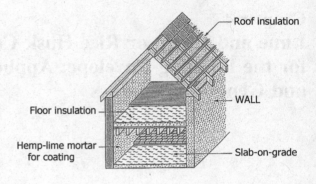

Table 4.1 Main characteristics and minimum required mechanical performances of hemp concrete on samples cured at 20 °C and 50%RH [2, 4, 8]

Application	Wall	Roof	Floor
Shiv content (wt%)	15	25	11
Bulk density (kg m^{-3})	400	250	500
Dry thermal conductivity (W m^{-1} K^{-1})	0.1	0.06	0.12
Minimum compressive strength (MPa)[a]	0.2	0.05	0.3
Minimum elastic modulus (MPa)[a]	15	3	15

[a]After 60 days of curing at 20 °C and 50%RH

For LHC cast on the building site, the structural studwork frame is encapsulated by hemp-lime. The studs can be positioned in either the center of the wall or on the inside edge. In this latter case, permanent shuttering against one face of the wall can be used [5].

Precast elements (less common and currently under development) are also used in conjunction with a structural studwork. However, it will be seen that some of them could be used as load-bearing bricks for single-storey houses.

4.1.2 Mix Design of Bio-Based Concretes

4.1.2.1 Manufacturing of Specimens

Manual Tamping

Rice husk and hemp shiv were first prewetted in a mixing drum during 5 min. The prewetting water (W$_P$) was chosen according to the water absorption test presented in Fig. 2.12. The lime-based binder (NHL3.5/CL90–S as presented in Table 3.5) was added in a second time and the mixing water was finally introduced.

The first stage of water uptake with a fast kinetics was considered to set the amount of prewetting water. It corresponds to 100 wt% for rice husk and 200 wt%

Fig. 4.2 a Hemp concrete cast on the building site. **b** Precast block of hemp concrete.
c Manufacturing by a projection process

for hemp shiv after 5 min (Fig. 2.12). Thereafter, the mixing water-to-binder mass
ratio (W_M/B) was taken as 0.5 and the water-to-binder mass ratio (W/B) was
calculated as follows (4.1 and 4.2):

$$LHC : W = W_P + W_M = 2A + 0.5B \tag{4.1}$$

$$LRC : W = W_P + W_M = A + 0.5B \tag{4.2}$$

where LHC and LRC are respectively Lime and Hemp Concrete or Lime and Rice
husk Concrete, W is the total water content, A is the aggregate content and B is the
binder content.

The fresh mixture was put into cylindrical $\Phi 11 \times 22$ cm^3 molds and compacted
in 3 layers using a steel manual device (Fig. 4.3). The height of a single layer is
equal to one-third the total height of the specimen (22 cm) and the mass of a single
layer is equal to one-third the total mass desired for the specimen, this mass being
calculated according to the target density of the freshly-mixed concrete.

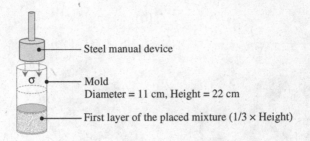

Steel manual device

Mold
Diameter = 11 cm, Height = 22 cm

First layer of the placed mixture (1/3 × Height)

Fig. 4.3 Compaction process of the mixture in the mold

Vibro-Compaction

A vibro-compression device (VCEC from MLPC®) initially dedicated to the compaction of soils and pavement materials was used in this case (Fig. 4.4).

The fresh mixes were vibro-compacted in cylindrical $\Phi10 \times 20$ cm^3 specimens.

The frequency of the pneumatic vibrator was 250 Hz according to the manufacturer.

The pneumatic cylinder was moved down in order to apply an axial compression (through a piston) and forced vibration in a perpendicular plane to the compression

Fig. 4.4 Vibro-compression device

axis was applied at the same time. This combined action of compression and vibration helps to reach the desired density by reducing the volume of voids due to the rearrangement of the granular skeleton. The mass of material introduced in the mold was chosen according to the target density of the material. The vibro-compression was stopped when the moving platen of the cylinder was in contact with the upper edge of the mold. After this stage, the material was unloaded.

According to Tronet et al. [6, 7], the water absorbed by plant aggregates is partly released during the compaction process. Hence, there is a need to lower the water-to-binder mass ratio in comparison to manual tamping for which the compaction pressure is low. This will be addressed afterwards. However, with this change, it should be noted that plant aggregates and one-third the amount of water (i.e., prewetting water) were mixed during 5 min. Then, the binder (Tradical® PF70) was introduced and 2 min later, the remainder of water was added.

4.1.2.2 Mix Proportions

Mix proportions of LHC (Lime and Hemp Concrete) and LRC (Lime and Rice husk Concrete) are given in Table 4.2. For plant-based concretes cast by manual tamping, the binder-to-aggregate mass ratio (B/A) ranged from 1.5 to 2.5. The mechanical behavior will be investigated on the mix for which B/A = 2 (Wall mixture). As regards vibro-compacted plant-based concretes, the binder-to-aggregate mass ratio was 2.3 as in the Ph.D. of Dinh where LHC is compacted with the same device [8].

Due to the weaker water absorption by rice husk, the water-to-binder mass ratio (W/B) of LRC was lower than that of LHC for a given B/A mass ratio. Furthermore, it can be seen that fresh density of LRC was higher than that of LHC for a given B/A mass ratio. This is related to the different particle density of rice

Table 4.2 Mix proportions and fresh density of hemp concrete (LHC) and rice husk concrete (LRC)

Concrete	CP[a]	B/A	W/B	A	B	W	FD[d]
				$kg\ m^{-3}$			
LHC	MT[b]	1.5	1.8	135	200	370	705
		2	1.5	145	285	430	860
		2.5	1.3	155	390	510	1055
	VC[c]	2.3	0.8	190	435	350	975
LRC	MT[b]	1.5	1.2	190	280	330	800
		2	1	195	395	390	980
		2.5	0.9	195	490	440	1125
	VC[c]	2.3	0.8	190	435	350	975

[a]Casting process
[b]Manual tamping on mixes using the binder NHL3.5/CL90–S
[c]Vibro-compaction on mixes using the binder PF70
[d]Fresh density

husk which is more than twice that of hemp shiv (650 kg m^{-3} for rice husk vs. 260 kg m^{-3} for hemp shiv). This higher apparent density of rice husk leads to a higher intergranular porosity (Table 2.2). Therefore, in order to ensure a minimum strength for LRC in the hardened state, the macroscopic intergranular porosity has to be limited. To achieve such a result, aggregate and binder contents were higher in rice husk-based mixes (Table 4.2).

In the case of the vibro-compacted mixture, the shiv content was higher than that used for manually tamped LHC so as to increase the compactness of the concrete after the casting process. Despite the different apparent density of the two kinds of aggregates, same mix proportions and same fresh density were targeted for LHC and LRC. This means that the rice husk content is finally very close to that reported for the manual tamping (i.e., 190–195 kg m^{-3}). Hence, the compaction intensity to cast LRC will be obviously lower to reach the design density of 975 kg m^{-3}. The W/B mass ratio was lowered to 0.8 for LHC as recommended by Dinh. The same ratio was adopted for LRC but it could be optimized.

The shiv content was 145 kg m^{-3} (17 wt%) for the mix with a B/A of 2 (typical ratio when LHC is used in timber frame walls). For LRC, the rice husk content was around 190 kg m^{-3} for all mixes. It ranged from 17.3 to 23.8 wt% with the decrease in the binder content.

4.1.2.3 Drying Kinetics

The cumulative mass loss of bio-based concretes until their hydric stabilization with the ambient air at 50%RH (for manual tamping) or 65%RH (for vibro-compaction) is reported in Fig. 4.5. Results are presented only for a B/A of 2 for manual tamping.

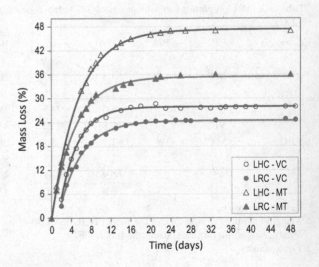

Fig. 4.5 Drying kinetics of specimens. *MT* Manual tamping (20 °C–50%RH, NHL3.5/CL90–S, B/A = 2), *VC* Vibro-compaction (20 °C–65%RH, PF70, B/A = 2.3)

For manually tamped samples, hydric stabilization is obtained for about 47% mass loss for LHC and 36% for LRC. As found by some authors [9, 10], about 90% of the initial water introduced in the material is discharged during the hydric stabilization process at room temperature and RH lower than 65%. This is the case if the mass loss is expressed in relation to the initial water content. The different mass loss for LHC and LRC is due to the different W/B (1.5 for LHC and 1 for LRC).

For vibro-compacted specimens, the final mass loss at 30 days is much lower. It is effectively 24% for LRC and 28% for LHC. The first explanation is obviously the W/B mass ratio lowered from 1.5 to 0.8 for LHC and from 1 to 0.8 for LRC. Nevertheless, if the mass loss is expressed in relation to the initial water content, it is noticed that the amount of evaporated water is lower (under 80%) than that calculated for samples cast by manual tamping (about 90%). This is probably due to the higher hydration degree of the PF70 binder compared to that of the NHL3.5/CL90–S binder. In addition, curing conditions could have an impact on this result since 65%RH is more favorable to hydration than 50%RH.

Drying kinetics of plant-based concretes depends upon the W/B mass ratio, the curing conditions, the type of binder used and the compaction degree (i.e., the degree of porosity). The latter will affect the carbonation kinetics. The curing period is characterized by an overlapping of drying (weight loss) and incipient carbonation which is known to be responsible for a significant weight increase. Since the carbonation process is controlled by CO_2 diffusion in the pores, it performs together with the drying process (according to the Fick's law). A different rate of initial carbonation (in relation to the rate of porosity and the pore size distribution) could explain the difference between LRC and LHC cast by vibro-compaction for which the W/B mass ratio is identical. Lastly, care must be taken with the desorption kinetics of plant aggregates. According to Fig. 2.13 (Chap. 2), the desorption behavior of hemp shiv from water saturation to 65%RH is different from that of rice husk.

4.1.2.4 Relations Between Density and Design Parameters

At lab-scale, many studies have focused on the properties of LHC with different mix proportions and target densities. Casting process, binder-to-aggregate (B/A) and water-to-binder (W/B) mass ratios are design parameters on which properties of LHC strongly depend.

Figure 4.6 reports the bulk density of various hemp concretes as a function of B/A.

The most widely used binder is the formulated lime Tradical® PF70 (made up of hydrated lime, hydraulic binder and pozzolan). The increase in bulk density with the decrease in the shiv content (increasing B/A) is clearly visible and follows a linear trend. This applies for manually tamped hemp-lime mixes. Lots of them are studied for B/A around 2 and a bulk density from 400 to 500 kg m^{-3}.

However, some mixes exhibit high densities for a rather low B/A and deviate from the trend. They correspond to a different casting process using static loading or

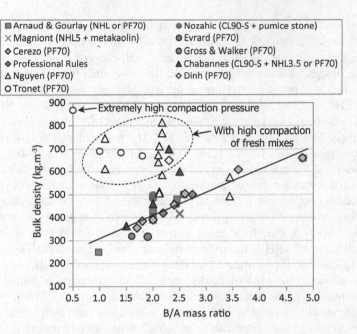

Fig. 4.6 Bulk density of hemp concretes as a function of B/A mass ratio [1, 3, 4–11, 16, 49]

vibro-compaction of the fresh mix. In fact, this process leads to a reduction of macroscopic intergranular voids and better arrangement of aggregates. It can applies for mixes with a B/A lower than 2.5 since it is related to the compaction ability of plant-derived aggregates. Hence, the compaction pressure can be increased gradually with the decrease in the B/A mass ratio. In this way, bulk density of LHC remains below 800 kg m^{-3} (except for one mix studied by Tronet [7] for which compaction pressure goes up to 7 MPa) (Fig. 4.6). For mixes of Nguyen [11] and Tronet [7], the compaction stress is applied until 48 or even 72 h of curing.

Figure 4.7 reports the bulk density of LRC as a function of B/A in comparison with LHC (mix proportions in Table 4.2) [3].

It can be seen that for a same B/A mass ratio, the bulk density of LRC is higher than that of LHC. This is due to the different particle density leading us to manufacture LRC with a higher fresh density in the case of manual tamping (Table 4.2).

After the vibro-compaction process, the bulk density of LHC is slightly less than 700 kg m^{-3} for a B/A mass ratio of 2.3. In this way, it is possible to have a close bulk density after hardening for both plant-based concretes. The rice husk content should be higher (>190 kg m^{-3}) to enhance the compaction of the granular skeleton in LRC. In this case, the reduction of the binder content is recommended in such a way to keep the bulk density under 800 kg m^{-3}.

Figure 4.8 reports the bulk density of hemp concretes as a function of the W/B mass ratio.

Fig. 4.7 Bulk density of LRC as a function of B/A mass ratio in comparison with LHC

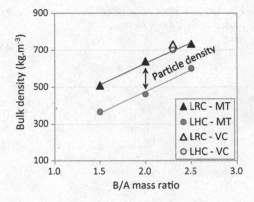

Fig. 4.8 Bulk density of LHC as a function of W/B mass ratio [1, 3, 4–11, 16, 49]

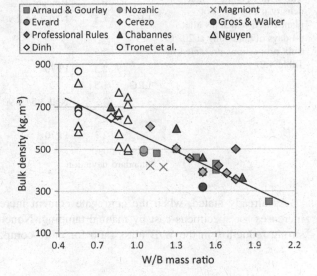

The W/B mass ratio of LHC (the water content) depends upon the aggregate content since the water absorption of plant particles is significant. Hence, for manually tamped LHC with a low density (i.e., low B/A), the water content is high. It decreases with the increase in the binder content (that is to say in the bulk density). Nevertheless, for LHC cast with a high compaction pressure of fresh mixes, the W/B mass ratio has to be relatively low (W/B < 1) since the water absorbed by plant aggregates is partially released during the compaction process. For highly compacted hemp-lime mixes, Tronet et al. [7] consider that a W/B of 0.55 is adequate. They assume that water for the wetting and the hydration of lime is entirely available at the end of compaction and they conclude that the water content has to be calculated as a function of the binder content.

To compare LRC with LHC, it is more interesting to plot the B/A mass ratio against the W/B mass ratio (Fig. 4.9).

Fig. 4.9 B/A mass ratio plotted against the W/B mass ratio

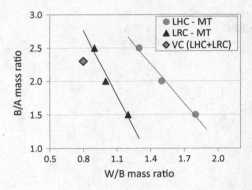

Table 4.3 Average bulk density (in kg m^{-3}) of plant-based concretes after 60 days of hardening at 20 °C (mix proportions presented in Table 4.2)

	B/A	Binder	%RH	MV[a]	Σ[b]
LHC	1.5	NHL3.5/CL90–S	50	364	5
	2			459	5
	2.5			600	14
	2.3	PF70	65	697	4
LRC	1.5		50	509	6
	2	NHL3.5/CL90–S		637	2
	2.5			734	12
	2.3	PF70	65	727	6

[a]Mean value
[b]Standard deviation

As already stated, when the aggregate content increases, the W/B mass ratio increases for specimens cast by manual tamping. Nonetheless, Fig. 4.9 shows the strong reduction in the W/B mass ratio for vibro-compacted LHC. For LRC, this

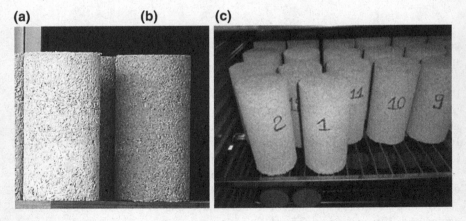

Fig. 4.10 Plant-based concretes after manual tamping. **a** LHC. **b** LRC. **c** Vibro-compacted LRC

reduction is much more moderate. The rice husk content and the fresh density are effectively very close to those of LRC cast by manual tamping with a B/A of 2 (A = 190–195 kg m^{-3} and fresh density of 975–980 kg m^{-3}). However, the W/B mass ratio of LRC using the PF70 binder could be probably further reduced.

Bulk densities of plant-based concretes are detailed in Table 4.3.

Specimens are pictured in Fig. 4.10.

4.2 Measuring Thermal and Mechanical Properties

4.2.1 Thermal Conductivity

The thermal conductivity of bio-based concretes is often measured using either the guarded hot plate method or the hot wire method.

4.2.1.1 Guarded Hot Plate

The guarded hot plate (ISO 8302 and EN 1946-2 standards) is the most commonly used and the most effective steady-state method for measuring the thermal conductivity of insulating materials [12]. It aims to establish a steady-state temperature gradient through a given specimen. The basic design of a guarded hot plate apparatus using a double-sided symmetric arrangement is represented in Fig. 4.11. In this model, two specimens of the same material and same thickness (d) are placed on each surface of the hot plate assembly and clamped by the cold plates. The meter plate is used to generate the heat flow (electrical power provided by Joule effect) in order to maintain the desired temperature gradient across the specimen (ΔT) [13]. Due to the finite dimensions of the specimens, the unidirectional heat flow is achieved through the use of guard heaters. As it can be seen in Fig. 4.11, the meter plate (power input) is surrounded by the guard plates with a gap between them. According to Yüksel [12], the cold plates are Peltier coolers or liquid-cooled heat sinks. Temperatures at the interface between the specimen and the plates are monitored by thermocouples. A well-defined and user-selectable temperature

Fig. 4.11 Basic design of a guarded hot plate system

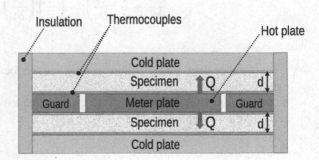

difference is established between the hot and the cold plates. The power input in the heater plate is measured as soon as thermal equilibrium is reached at steady-state conditions. A data acquisition system is connected to the temperature and the electrical power supply devices controlled in turn by a closed-loop system [12]. In fact, the temperature of the meter plate is changeable by adjusting the input power of the electrical heater embedded within the meter plate [13]. The thermal conductivity of the specimen is calculated based on the Fourier's law by the temperature difference between the meter and cold plates (ΔT), the heat input to the meter plate (Q), the measurement surface area (A) and the thickness of the specimens (d) (4.3):

$$\lambda = \frac{Q}{2} \times \frac{d}{A.\Delta T} \qquad (4.3)$$

where λ is the thermal conductivity (W m^{-1} K^{-1}), Q is the heat power (W), ΔT (i.e., $T_{hot} - T_{cold}$) is the temperature differential across the specimen (Kelvin), d is the specimen thickness (m) and A is the heat transfer area (m^2).

This equation is valuable for a double-sided apparatus where Q/2 goes upward and the other half goes downward (Fig. 4.11).

The disadvantage of this method is that establishing a steady-state temperature gradient through the specimen is time-consuming [12].

4.2.1.2 Hot-Wire Method

This method is a transient technique (NF EN ISO 8894 standard) based on recording the rise in temperature at a defined distance from the heat source [12]. The model used assumes a unidirectional heat transfer. The heat flow (linked to the electrical power delivered by the apparatus) and the temperature rise are simultaneously measured. The electrical power (P) is constant and the temperature rise in the hot wire is measured by a welded thermocouple.

Considering the hot-wire probe as an ideal infinitely thin and long line heating source and the studied material as an homogeneous and isotropic one, the heat conduction equation in cylindrical coordinates is written as follows (4.4) [14, 15]:

$$\frac{\partial^2 T}{\partial r^2} + \frac{1}{r}\left(\frac{\partial T}{\partial r}\right) = \left(\frac{1}{\alpha}\right)\left(\frac{\partial T}{\partial t}\right) \qquad (4.4)$$

with:

- r, the distance between the heating source and the location where temperature is measured (m)
- α, the thermal diffusivity (m^2 s^{-1}) which is correlated with thermal conductivity according to the following Eq. (4.5):

$$\propto = \frac{\lambda}{\rho \times C_M} \tag{4.5}$$

where λ is thermal conductivity (W m^{-1} K^{-1}), ρ is the bulk density of the material and C_M is the specific heat (J K^{-1} kg^{-1}).

Based on 4.4, for a sufficient long time, there is a proportional relationship between temperature rise (ΔT) and logarithmic heating time [i.e., ln(t)] [14]. The following Eq. (4.6) can be written if the temperature is measured at times t_1 and t_2:

$$\lambda = \frac{P}{4\pi L \times \Delta T} \ln(t_2/t_1) \tag{4.6}$$

where P is the electrical power (W), ΔT is the temperature variation (Kelvin), L is the wire length (m) and λ is thermal conductivity (W m^{-1} K^{-1}).

The hot wire is situated between two equally sized homogeneous specimens as shown in Fig. 4.12b with two cylindrical specimens. Thermal conductivity is calculated by comparing the plot of the wire temperature versus the logarithm of time. The typical thermogram representing ΔT as a function of ln(t) is represented in Fig. 4.12a. Actually, thermal conductivity is calculated using the linear portion (long-term slope ζ) and it is expressed as follows (4.7):

$$\lambda = \frac{P}{4\pi L} \times \frac{\Delta \ln(t)}{\Delta T} = \frac{P}{4\pi L\zeta} \tag{4.7}$$

Thermal conductivity of manually tamped LHC and LRC studied by the authors was measured using the hot-wire method. The device was a commercial CT-meter (NF EN 993-15 standard). The rise in temperature measured by the sensor was limited to 20 °C, the heating time was taken as 400 s and the power supply was 0.2 W.

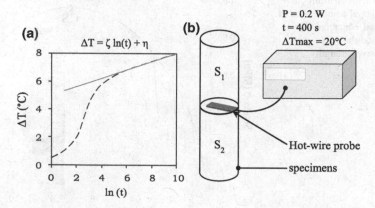

Fig. 4.12 a Hot-wire typical thermogram. b Thermal conductivity measurement

This transient method presents two main advantages. Compared to the steady-state guarded hot plate, test time is shorter. Moreover, it is entirely possible to study the effect of moisture content on thermal conductivity (for instance measurement at 50%RH of after drying). Indeed, according to Collet and Pretot [14], a transient method does not induce water migration during the test. However, it should be noted that measurements are localized.

4.2.2 Mechanical Properties Under Uniaxial Compression

Mechanical characteristics in compression (compressive strength, Young's modulus) were determined with a standard testing machine. Cylindrical specimens ($\Phi 11 \times 22$ or $\Phi 10 \times 20$ cm^3) were tested with a loading rate of 5 mm min^{-1} (displacement-controlled testing) using a spherical seat on the upper plate since top and bottom surfaces of specimens are not necessarily parallel. Cycles of loading/disloading were performed at 1, 2 and 3% strain so that to determine the tangent modulus (E_C) on the loading phase. In view of the rigidity of the testing frame which is largely higher than that of specimens, the strain was easily determined by using the displacement of the frame and the initial height of the specimen. The stress-strain curve of a plant-based concrete is reported in Fig. 4.13.

At the beginning of the test ($\varepsilon < 1\%$), the mechanical behavior of plant-based concretes is considered to be quasi-elastic. Some authors define the elastic modulus by referring to the initial slope of stress-strain curves [4, 16]. According to Chamoin [17], it is far better to define it on the loading cycles (E_C in Fig. 4.13). After a given strain, plant-based concretes (both LHC and LRC) change their mechanical behavior towards an elastoplastic one. During the phase corresponding to the quasi-elastic linear part of the curve, the binder fully supports the compressive

Fig. 4.13 Stress-strain curve under uniaxial compression of plant-based concretes (E_C Elastic modulus, CS Compressive strength)

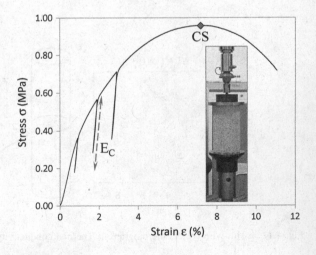

stresses. When the behavior becomes elastoplastic, interfaces between the binder and the plant aggregates are progressively damaged until the load is largely transferred to the plant aggregates. Under high deformations, the stress keeps going up without reaching complete failure because of the compaction ability of aggregates due to their significant deformability.

4.2.3 Shear Strength

4.2.3.1 Triaxial Shear Test on Plant-Based Concretes: Experimental Device and Test Procedure

The shear behavior of bio-based concretes (LHC and LRC) was investigated with an adapted triaxial apparatus. Figure 4.14 shows a schematic representation of the triaxial cell with the specific experimental conditions used to study the shearing of bio-based concretes ($\Phi10 \times 20$ cm^3). The equipment is composed of a load frame with a measuring device of the axial force (load sensor of 50 kN with high accuracy of measurement), a triaxial cell that comes to be fixed on a speed-controlled platen with a piston that transmits the axial load to the specimen, a large displacement transducer located on the moving platen (to measure the axial strain) and a cell pressure controller to set the confining pressure σ'_3 in the cell. The temperature was maintained at 20 °C in the room. All data (load cell, axial strain and cell pressure) were collected by the GDS Lab acquisition software.

Specimens were put within latex membranes in 0.3 mm thickness and sealing was guaranteed with rubber O-rings. A porous stone was placed between the bottom of specimens and the moving platen and the heads of specimens were surfaced by a thin layer of aluminous cement (Fig. 4.14). The unsaturated specimens were drained at air pressure by holding the valves open as shown in Fig. 4.14.

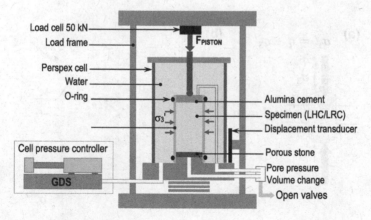

Fig. 4.14 Schematic representation of the triaxial device

In these conditions, the actual configuration of a plant-based concrete wall is simulated.

Displacement-controlled tests were performed with a speed rate of 0.4 mm min^{-1}. As for unconfined compression, a loading/disloading cycle was performed at 2% strain in order to estimate Young's moduli. The triaxial test was performed after 60 days of curing on specimens cast by the vibro-compaction method previously described.

Several initial effective confining pressures (p'_0 = 25, 50, 100 and 150 kPa) were applied and 3 or even 4 specimens were tested for each effective confining pressure p'_0.

4.2.3.2   Evaluation of Shear Strength Parameters in the Cambridge Diagram (q − σ'_m)

In a conventional triaxial test, principal stresses are σ'_1, σ'_2 and σ'_3 with $\sigma'_2 = \sigma'_3$ (Fig. 4.15a). The mean effective pressure (noted σ'_m) is defined as follows (4.8):

$$\sigma'_m = \frac{\sigma'_1 + 2\sigma'_3}{3} \tag{4.8}$$

where σ'_1 is the effective axial stress and σ'_3 is the effective confining stress (Fig. 4.15a). The deviatoric stress ($q = \sigma'_1 - \sigma'_3$) can therefore be expressed as a function of the mean effective pressure in the following manner (4.9):

$$\sigma'_m = \frac{q}{3} + \sigma'_3 \tag{4.9}$$

The ratio between the deviatoric stress and the mean effective pressure (called stress ratio) represents the mobilized stresses for a given loading path. The latter can be

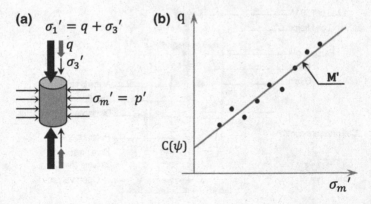

Fig. 4.15   a Stress state during triaxial compression. **b** Failure line in coordinates q − σ'_m

determined for the peak deviatoric stress or for higher strains. It corresponds to the slope of the failure line in the $q - \sigma'_m$ diagram $(M' = q/\sigma'_m)$ (Fig. 4.15b). It is then possible to calculate the friction angle φ by the following relation (4.10): [18, 19]

$$\varphi = \arcsin\left(\frac{3M'}{M'+6}\right) \tag{4.10}$$

Thereafter, the cohesion (noted C) of the material is expressed as follows (4.11):

$$C = C(\psi) \times \left(\frac{\sin \varphi}{M'}\right) \tag{4.11}$$

where $C(\psi)$ is the intercept of the linear regression reported in Fig. 4.15b, φ is the friction angle previously calculated and M' is the effective stress ratio.

4.3 From Design to Specific Properties of Bio-Based Concretes

4.3.1 Porosity and Thermal Conductivity

Hemp concrete features interesting hygrothermal properties owing to its high porosity [17]. The porosity of plant-based concretes can be described as a triple porosity with interconnections between the pores [20] (Fig. 4.16):

- macroscopic inter-particles porosity (millimetric)
- porosity within the particles (~ 10 μm for hemp shiv)
- binder porosity (from 10 nm to 1 μm depending on the binder).

Fig. 4.16 Triple porosity of LHC. **a** Macroscopic inter-particles porosity. **b** Porosity within hemp shiv. **c** Porosity of the binder

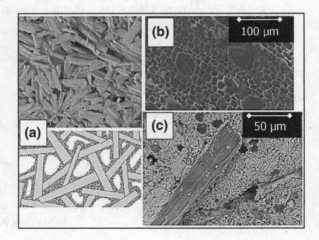

According to Collet et al. [21], the total porosity of hemp concrete with a bulk density around 400 kg m^{-3} is between 76 and 78% with an open porosity of about 70%. This high porosity is responsible for a low thermal conductivity due to the big amount of trapped air but it also ensures a very good capacity to store and release moisture and to moderate sudden step change of the indoor moisture level. Indeed, porous materials easily trap water molecules in moist air and release moisture when air becomes dry. This is a fundamental characteristic for exchange of water vapor inside and outside the buildings [10].

4.3.1.1 Porosity Estimation

The inter-particles porosity into the hardened plant-based concrete can be calculated according to the following Eq. (4.12):

$$\eta_{IP} = \frac{(M_A/\rho_A) + [(M_B \times t)/\rho_{SP}]}{V_S} \tag{4.12}$$

where M_A is the mass of aggregates introduced in the specimen, ρ_A is the apparent density of the particles, M_B is the mass of binder, t is the hydration degree of lime-based binders (taken as 1.1 according to Tronet et al. [7]), ρ_{SP} is the specific density of the binder and V_S is the volume of the specimen (i.e., of the mold).

The total porosity (including capillary voids within plant particles) is calculated in the same manner by replacing the apparent density (ρ_A) by the true density of particles (ρ_T) in (4.12).

These porosities are plotted against the bulk density of plant-based concretes in Fig. 4.17. From the latter, it is seen that the inter-particles porosity of LRC is significantly higher than that of LHC for a given bulk density. This is due to the higher apparent density of rice husk leading to an important inter-granular porosity. Moreover, the monodisperse granulometric distribution of rice husk is likely to generate more intergranular voids. Furthermore, the total porosity of plant-based concretes for a given bulk density is very close. This is due to the high porosity

Fig. 4.17 Porosity (*IP* Inter-particles porosity, *TOT* Total porosity) plotted against bulk density of plant-based concretes

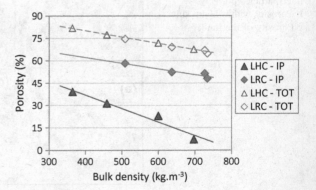

Fig. 4.18 Porosity plotted against B/A (*MT* Manual tamping, *VC* Vibro-compaction)

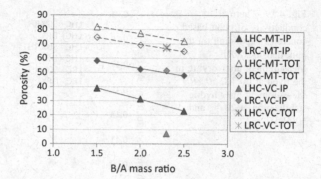

within hemp shiv and this is in accordance with the total porosity in bulk aggregates reported in Table 2.2. For a given bulk density, the B/A mass ratio of LRC is lower than that of LHC. Therefore, it is also interesting to plot the porosity against B/A (Fig. 4.18).

From Fig. 4.18, it can be drawn that the inter-particles porosity of LRC is higher than that of LHC for a given B/A as was the case with bulk density. For instance, the inter-particles porosity of LRC with a B/A of 2 is more than 50% whereas it is barely above 30% for LHC even though the bulk density of LRC is higher (due to higher binder and aggregate contents). This shows that the high amount of inter-granular voids in rice husk aggregates is clearly responsible for this high inter-particles porosity in LRC. For plant-based concretes cast by manual tamping, the inter-particles porosity decreases with the B/A mass ratio (and with the bulk density). For LRC, this is due to the increase in the binder content (since the aggregate content is 190-195 kg m^{-3} for all the mixes). Thus, the binder can fill the voids between the aggregates. For LHC, the increase in the B/A is also the result of the increase in the binder content. However, the hemp shiv content also increases (from 135 to 155 kg m^{-3}) (Table 4.2). This leads to a higher compaction stress to achieve the desired density. This explains why the decrease in the inter-particles porosity as a function of B/A is more significant for LHC (Fig. 4.18). Moreover, the very low inter-particles porosity of vibro-compacted LHC is highlighted (7.2%). This is due to the higher binder content but also to the compaction of hemp shives introduced in greater amount (Table 4.2). As regards total porosity, it is slightly lower for LRC. This is due to its higher density. When the bulk density is the same, as is the case after the vibro-compaction process, the total porosity is equivalent.

4.3.1.2   Thermal Conductivity of Manually Tamped Specimens

Thermal conductivity of plant-based concretes depending on the bulk density is reported in Fig. 4.19. It was measured by the hot-wire method for the three mixes of LHC and LRC with a B/A mass ratio varying from 1.5 to 2.5 (for which porosities are calculated and presented in the previous section). Results are also compared with measurements from literature (Fig. 4.19) [14, 10, 16].

Fig. 4.19 Thermal conductivity of plant-based concretes depending on the bulk density (*DRY* Measurement after drying in the oven at 60 °C, *50%RH* Measurement after hydric stabilization at 20 °C and 50%RH)

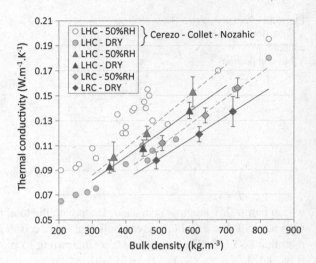

Thermal conductivity depends on binder and aggregate contents. It increases linearly with the bulk density (i.e., increase in the binder content or in the compaction pressure). It is also linked to the water content in the plant-based concrete. As it can be seen in Fig. 4.19, thermal conductivity measured after the drying of LHC and LRC (48 h at 60 °C) is lower than that measured after the hydric stabilization of specimens at 20 °C and 50%RH. Furthermore, a relative dispersion of results is noted for literature measurements on LHC after 50%RH. This is due to the different method used for the measurement (hot-wire method or guarded hot plate). It should be remembered that the guarded hot plate method is not necessary accurate for high moisture contents within the specimens.

For a same density, LRC shows a lower thermal conductivity. The comparison of this graph with the evolution of porosity as a function of density shows that the macroscopic inter-particles porosity plays a key role in thermal performances. The thermal conductivity of LHC and LRC is compared from the B/A point of view in Fig. 4.20.

It appears that both plant-based concretes present a roughly equivalent thermal conductivity. This shows that the B/A mass ratio is closely related to the thermal performances. For a given B/A, LRC is denser than LHC but its thermal conductivity remains low. This confirms that the high amount of intergranular voids in bulk rice husks is such that LRC can compete with LHC in terms of thermal conductivity.

4.3.1.3   Thermal Conductivity of Precast Blocks

In most cases, the thermal conductivity of plant-based concretes is measured in the parallel direction to the compaction axis (lozenge-shaped pattern in Fig. 4.21).

Fig. 4.20 Thermal conductivity of plant-based concretes depending on the B/A mass ratio

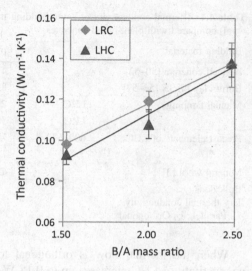

Fig. 4.21 Dry thermal conductivity of LHC as a function of bulk density for manual tamping and compacted blocks. *Perp* Perpendicular ($\lambda_\perp$), *Par* Parallel ($\lambda_{//}$)

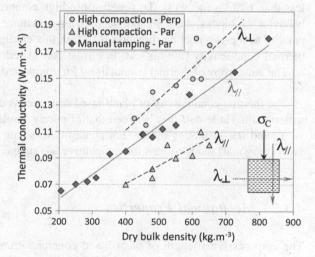

However, Nguyen [11] also measured the thermal conductivity of compacted blocks in the perpendicular direction to the compaction axis. The thermal conductivity measured in the parallel direction to the compaction axis for compacted blocks is more favorable than that measured on specimens cast by manual tamping. By contrast, the thermal conductivity of LHC measured in the perpendicular direction to the compaction axis is higher (i.e., less favorable). This is due to the anisotropic behavior. The compaction of LHC generates an anisotropic layered composite with preferentially oriented particles in the perpendicular direction to the compaction axis. The heat flow is favored when it takes place in the same direction than that of longitudinal capillaries of shiv and inversely.

Table 4.4 Thermal conductivity of some building materials (self-insulating blocks and mineral wool) compared with plant-based concretes

Buiding material		ρ^a (kg m^{-3})	λ^b_{dry} (W m^{-1} K^{-1})
Aerated concrete [50–53]		415–600	0.12–0.16
Burnt clay bricks [54–57]		850–1780	0.24–0.96
Manual tamping	LHC	250–500	0.08–0.11
	LRC	500–730	0.10–0.14
Precast elements of LHC	600–670	λ^c_P	0.10–0.13
		λ^c_O	0.14–0.18
Mineral wool [3]		15–40	0.03–0.045

^{a}Bulk density
^{b}Dry thermal conductivity
$^c\lambda_P$ Parallel, λ_O Orthogonal

When the thermal flow is orthogonal to the compaction direction, thermal conductivity of LHC increases up to 0.18 W m^{-1} K^{-1} for a bulk density which is less than 670 kg m^{-3} [11]. To benefit from high compressive strength and ductile behavior of blocks, they should be used in a manner that compression load is parallel to the compaction direction [6, 7]. In this configuration, the perpendicular thermal conductivity (the highest) has to be considered.

The anisotropy of thermal conductivity for compacted LRC has not been studied yet.

The thermal conductivity of plant-based concretes is compared to other building materials in Table 4.4. It is between that of purely insulating materials (as mineral wool) and self-insulating blocks. For precast blocks with high compaction pressure, λ_O (orthogonal) tends to lose its competitive advantage on cellular blocks.

4.3.2 Mechanical Properties

The compressive strength of plant-based concretes from literature is presented in Fig. 4.22. It is plotted against the bulk density.

The mechanical performances of bio-based concretes depend upon many factors as the type of binder (only air lime with varying proportions of additional hydraulic and pozzolanic admixtures is considered in this book), the binder content, the casting process (manual tamping or high-pressure compaction of the material), the properties of plant particles (including water absorption, chemical bonding or size distribution), the age of the concrete and curing conditions (temperature, relative humidity, CO_2 concentration). The last two will be the subject of the next section. For hemp-lime mixes cast around a timber frame, the bulk density at dry state is between 300 and 500 kg m^{-3} for a large part of studies (Elfordy et al. [22], Arnaud and Gourlay [4], Cerezo [16]). The compressive strength of LHC in this configuration is 0.2–1 MPa for a density ranging from 250 to 800 kg m^{-3}. For a B/A mass

Fig. 4.22 Compressive Strength (CS) of plant-based concretes as a function of bulk density and influence of the casting process after 28 days (3 months for Cerezo), CS of Nguyen is given for 7.5% strain and for Tronet, refer to the yield stress defined in Tronet et al. [7] (① Manual tamping, ② Mechanical compaction at fresh state)

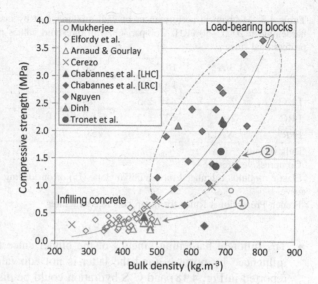

ratio around 2 (and density lower than 500 kg m^{-3}), the compressive strength does not exceed 0.5 MPa (Fig. 4.22). In fact, densities higher than 500 kg m^{-3} are obtained by increasing the binder content (B/A = 2.7 for Mukherjee [23] and 4.8 for one mix of Cerezo [16]). This method provides a compressive strength higher than 0.5 MPa (0.98 MPa for Cerezo [16]). Nevertheless, using a low aggregate content is detrimental to the thermal performances.

Results from Nguyen [11] and Tronet et al. [7] are also reported in Fig. 4.22 and are clearly distinguished from those of other authors. This is due to the high compaction of fresh mixes leading to strongly reduced macroscopic inter-particles porosity. When the compaction pressure is very high, the stress at 7.5% strain almost reaches 4 MPa. In the studies of Nguyen [11] and Tronet et al. [7], the compaction pressure is maintained during 48 or even 72 h of curing before demolding. This is possible by decreasing the binder content so that the bulk density remains under 800 kg m^{-3}. In addition, an important compaction pressure is possible when the shiv content is sufficiently high (due to the compressibility of particles). When the compaction pressure is high, the stress-strain behavior shows an increase in ductility (large strain hardening area) [7].

Table 4.5 reports the mechanical performances of wall mixtures cast by manual tamping for LHC and LRC (mix proportions in Table 4.2) after 60 days of hardening. Despite the higher bulk density of LRC, the latter exhibits lower mechanical performances. However, it is seen that compressive strength and elastic modulus of LRC are beyond the threshold values of the French professional rules [1] recommended for LHC.

At this stage, the reasons for the lower mechanical performances of LRC could be the following:

Table 4.5 Mechanical performances of wall mixtures cast by manual tamping after 60 days of hardening at 20 °C–50%RH. Comparison with threshold values recommended by LHC French professional rules

	B/A	BD[a]	CS[b]	E_I^c	E_C^c
		kg m^{-3}	MPa	MPa	MPa
LHC	2	459 ± 5	0.48 ± 0.02	27 ± 1	73 ± 3
LRC	2	637 ± 2	0.33 ± 0.03	17 ± 4	50 ± 5
FPR[d]	WALL		>0.2	>15	–

[a]Bulk density
[b]Compressive strength
[c]Elastic modulus calculated on the initial slope (E_I) or on loading cycles (E_C) of the stress-strain curve
[d]French Professional Rules for hemp concrete structures

- The different hardening kinetics of the binder since the carbonation process is influenced by porosity (and the latter is not equivalent for LRC and LHC as reported in Fig. 4.18) and C_2S hydration could be disturbed by the presence of polysaccharides. These parameters are directly linked to density, size distribution and chemical composition of plant aggregates.
- The weaker bond strength of rice husk with the lime-based binder (chemical bonding, mechanical interlocking, surface hydrophilicity, particle stiffness, specific morphology).
- The granular stacking and the mechanical response of the granular skeleton as the inter-particles porosity of LRC is much higher than that of LHC according to porosity estimations (which are in relation with physical properties of rice husk and poor rearrangement due to the size distribution).

The first point will be addressed in the next section that deals with the hardening kinetics depending on curing conditions.

As regards the influence of biomass aggregates on the hardening mechanisms of binders, some elements are reported below.

Setting and hardening mechanisms of the binder can be disturbed by polysaccharides extracted from the cell wall of particles. Sugars, pectins, hemicelluloses and carboxylic acids are known to hinder the hydration of calcium silicates (C_3S and C_2S), inducing a set retardation of hydraulic binders (especially Portland cement). For instance, pectins trap calcium ions (Ca^{2+}) and those are not available for dissolution-precipitation processes that operate during the hydration of calcium silicate binders [24, 25]. Diquelou et al. [26] studied the setting time and the compressive strength of cement pastes designed either with water or with a leach solution prepared by soaking hemp shiv in water during 24 h. The authors used hemp shives from different origins and for a given kind of hemp shiv, they found that soluble components are not only responsible for a setting delay but also for a lower compressive strength after 28 days compared to the neat cement paste. This result clearly underlines the disrupting effect of some polysaccharides on the hydration process. The same kind of investigation was conducted by Walker and

Fig. 4.23 Setting time of calcic lime (CL) and lime-pozzolan pastes with water or with the leach solution (LS). *CL-GGBS* Lime- Ground granulated blast-furnace slag. *CL-MK* Lime-Metakaolin. *CL-RHA* Lime-Rice husk ash

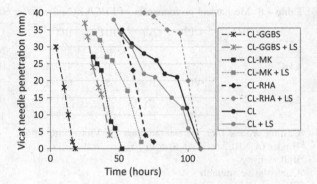

Pavia [25] using lime-pozzolan blends as binders. It has been shown that hemp does not alter the setting time of pure lime as it can be seen in Fig. 4.23 where the results of the Vicat needle test are reported.

In addition, it has been evidenced that hemp shiv does not affect the strength development of lime pastes (Fig. 4.24). According to the authors, the results indicate that hemp shiv does not alter flocculation, drying and early carbonation responsible for the initial hardening of aerial lime. However, a delay in the setting time is observed for lime-pozzolan pastes (Fig. 4.23). Hemp also delays the strength development of the pastes using supplementary cementing materials (metakaolin and ground granulated blast-furnace slag in Fig. 4.24). This confirms that hemp extracts delay the formation of pozzolanic C–S–H. Nevertheless, contrary to what has been reported for cement pastes, the ultimate compressive strength of lime-pozzolan pastes is not affected by the leach solution (Fig. 4.24).

Mechanical performances of LHC and LRC cast by manual tamping are compared with those achieved by the vibro-compacted specimens (mix proportions in Table 4.2) in Table 4.6.

Fig. 4.24 Compressive strength of calcic lime (CL) and lime-pozzolan pastes (*MK* Metakaolin, *GGBS* Ground granulated blast-furnace slag) with water or with the leach solution (LS)

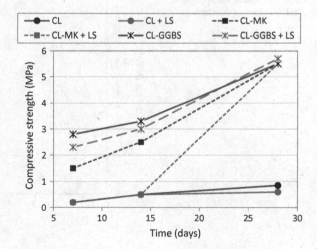

Table 4.6 Mechanical performances of LHC/LRC mixes after 60 days at 20 °C

	B/A	CP[a]	B[b]	RH	BD[c]	CS[d]	E_C^e
				%	kg m^{-3}	MPa	MPa
LHC	2	MT	A	50	459 ± 5	0.48 ± 0.02	73 ± 3
	2.3	VC	B	65	697 ± 4	2.47 ± 0.23	236 ± 28
LRC	2	MT	A	50	637 ± 2	0.33 ± 0.03	50 ± 5
	2.3	VC	B	65	727 ± 6	1.46 ± 0.18	175 ± 30

[a]Casting process (*MT* Manual tamping, *VC* Vibro-compaction)
[b]Binder (*A* NHL3.5/CL90–S, *B* PF70)
[c]Bulk density
[d]Compressive strength
[e]Elasic modulus

Once again, and despite identical mix proportions and close density, the compressive strength of the rice husk-based concrete is 1.7 times lower than that recorded for LHC. The granular stacking of plant-based concretes has a great chance to play a key role in this result as the inter-particles porosity of LHC is found to be only 7.2% whereas that of LRC is 51.3% (see in Fig. 4.18). For equivalent binder content, it can be assumed that the compressive strength of LHC and LRC with increasing aggregate content will follow a different trend.

In addition, it should be noted that the compressive strength of LRC–VC is 4.4 times higher than that of LRC–MT. This is not due to a higher compactness as the inter-particles porosity is around 52% in both cases. Consequently, this significant strength gain is attributed to the higher B/A mass ratio (2.3 against 2) and the higher degree of hydraulicity of the binder PF70 compared to NHL3.5/CL90–S. This

Fig. 4.25 Stress-strain curves of plant-based concretes under compression after 60 days of hardening

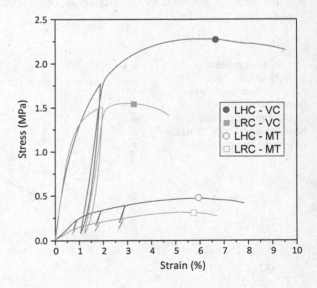

resulted in a higher bulk density for LRC–VC (730 kg m^{-3}) than that noted for LRC–MT (640 kg m^{-3}). Moreover, the relative humidity (65 vs. 50%RH) and the lower W/B mass ratio (Table 4.2) could have an interesting influence. As regards LHC, the effect of the increased compactness (inter-particles porosity of 7% after vibro-compression against 31% after manual tamping) obviously comes into play.

Stress-Strain curves for these mixes previously compared in Table 4.6 are presented in Fig. 4.25.

From Fig. 4.25, it can be seen that the strain at peak stress is equivalent for both plant-based concretes cast by manual tamping ($\sim$6% strain). However, LHC exhibits a more ductile behavior than LRC for vibro-compacted specimens. The strain at failure is more than 6% for LHC whereas it is only 3% for LRC. This confirms that the granular skeleton of hemp shives was much more compacted than that of rice husks. The closing of the inter-particles porosity of LHC in the fresh state results in the compression of a dense packing of shives with a high strain capacity in the hardened state. Furthermore, the ductility of vibro-compacted specimens is not enhanced compared to specimens cast by manual tamping. The trend is even towards brittleness for LRC (Fig. 4.25). This is certainly due to the higher binder content. This assumption confirms that the mechanical behavior of LHC after vibro-compaction results primarily from the reduction of intergranular voids whereas that of LRC is rather the result of the higher binder content.

4.4   Effect of Curing Conditions on Hardening and Mechanical Properties of Plant-Based Concretes

4.4.1   Curing Conditions

The main weakness of plant-based concretes using lime as binder is the long time they require to cure when cast in situ. Using accelerated carbonation appears promising with highly porous bio-based concretes designed with a large part of $Ca(OH)_2$. Some studies have dealt with the effect of accelerated carbonation on vegetable fiber reinforced composites. They are presented as good initiative to CO_2 sequestration and an interesting way to decrease alkalinity and porosity within concrete. A smaller average pore diameter associated with a densification of the matrix by higher precipitation of $CaCO_3$ results in increased bulk density, improved mechanical properties and enhanced durability [27, 28]. The study about accelerated carbonation is carried out in the prospect of moving towards a load-bearing material for single-storey houses using precast bricks without affecting thermal performances too much.

Under optimal conditions for carbonation (65%RH), Lanas et al. [29] have shown that C_2S hydration mainly occurs after 1 month in hydraulic lime mortars. Therefore, its contribution to strength is low before this date. This was confirmed in the section of the book about lime-based binders. However, hydration time of C_2S

Table 4.7 Mix proportions and fresh density of LHC and LRC for which the effect of curing conditions are investigated

Concrete	B/A	W/B	A	B	W	Fresh density
			kg m^{-3}			
LHC	2	1.5	145	285	430	860
LRC	2	1	195	395	390	980

depends on temperature and relative humidity. Indeed, it was shown that high RH (>95%) and elevated temperature strongly accelerate C_2S hydration at early age, resulting in the increase of compressive strength.

The following curing conditions and histories were studied:

① Indoor Standard Conditions (ISC) in a climate-controlled room at 20 °C and 50%RH during 10 months.
② Outdoor exposure Conditions (OC) with a recording of temperature and relative humidity under shelter during 10 months.
③ 40 days at 20 °C–50%RH followed by 1 month under accelerated carbonation curing (ACC) at 65%RH.
④ Moist curing in airtight enclosures at 20 °C–95%RH during 7 days or 21 days and standard curing (20 °C–50%RH) until 28 days (7d–MC or 21d–MC).
⑤ Thermal activation at 50 °C–95%RH during 7 days and standard curing (20 °C–50%RH) until 28 days (7d–TA).

①, ④ and ⑤ are the same curing histories as those used to study the hardening and mechanical performances of lime-based mortars (Chap. 3). The effect of curing conditions is investigated on LHC and LRC cast by manual tamping with the binder NHL3.5/CL90–S. Mix proportions and fresh densities are recalled in Table 4.7.

4.4.1.1 Outdoor Exposure Conditions (OC)

The profile of temperature and %RH obtained outdoors is reported in Fig. 4.26a.

The first period of curing corresponds to winter with an average temperature of 10 °C and a relative humidity remaining rather high. By going towards summer, temperature increased by up to 30 °C.

Important variations in relative humidity throughout the entire period can be drawn from the profile. An accurate analysis is presented in the box plot in Fig. 4.26b. From this, it can be concluded that RH mostly ranged between 45 and 75% during the outdoor exposure curing (OC). This range of values is conducive to carbonation. The influence of temperature is less known but tests have shown that more calcite is formed if cold carbonic acid is used for carbonation, in the range 0–10 °C [30].

Fig. 4.26 Data acquisition of temperature and RH during 10 months of outdoor exposure (OC). **a** RH and temperature profiles. **b** RH distribution during outdoor exposure

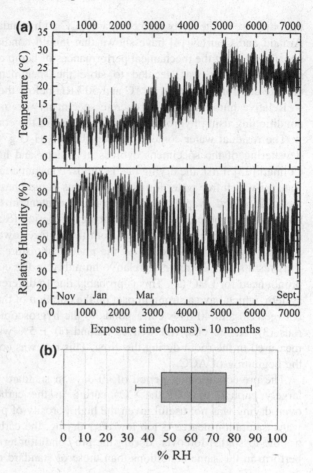

4.4.1.2 Accelerated Carbonation Curing (ACC)

Initial Conditioning Before ACC

Diffusion of CO_2 within concrete and chemical reaction kinetics are influenced by relative humidity in the environment and degree of saturation of pore spaces. The water content of the concrete just before the start of the carbonation process is an important factor as well as relative humidity in the gaseous mixture during the carbonation test. According to the French Standard XP P18-458, ordinary concretes are first cured under 100%RH during 28 days before being pre-conditioned in order to desaturate the pores on the surface and facilitate CO_2 diffusion. For this to happen, an oven-drying at 40 °C during 14 days is regularly used [31, 32]. Plant-based concretes cannot be cured under moist conditions during 1 month since the objective of the study is to compare ACC with standard conditions. In addition, Arnaud and Gourlay [4] have shown that humid conditions (RH > 75%) tend to negatively affect the mechanical performances of hemp concretes. Therefore, in the present study, it was decided to store the specimens during 40 days in the climate-controlled room at 20 °C and 50%RH before the beginning of CO_2 curing.

Relative humidity inside concrete specimens was measured during the initial conditioning using hygrometric probes placed in the core (Fig. 4.27).

The residual water content on a dry basis (gH_2O $g^{-1}dry$) is also reported. The dewatering of the specimens evolves in 3 steps and has been well described by Colinart [33]. First, the drying rate is almost constant and corresponds to a funicular state for which free water forms a continuous liquid phase and can be transported to the surface and evaporated. The moisture content in the core of the specimens remains high (more than 95%RH) during this period (Stage 1 in Fig. 4.27). After a few days, the drying front moves from the surface towards the core of specimens and the drying rate controlled by the internal moisture transfer decreases sharply. This results in a decrease of relative humidity in the core, which is slightly more pronounced for LHC (2). This is probably due to different topologies of the porous system, which have a strong impact on the duration of this second stage. The period between 30 and 40 days corresponds to the hygroscopic equilibrium of the materials (3). At the end, RH weaves around 60 ± 5% which is approximately that measured in the room during this time. This rate was considered as well suited for the beginning of ACC.

The pre-conditioning period of 40 days in standard conditions was considered largely enough to start the CO_2 curing in the carbonation chamber and the oven-drying was not useful given the high porosity of plant-based concretes. If the water saturation degree is too low after drying, the carbonation process may have trouble to start. Moreover, during the pre-conditioning period, C_2S hydration can perform in the same conditions than those of standard curing.

Fig. 4.27 Variation of RH and residual water content within specimens during the initial conditioning

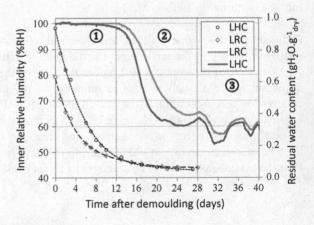

Accelerated Carbonation Curing (ACC)

After 40 days, LRC/LHC were introduced in glass enclosures fed by CO_2 during 30 days (see Fig. 4.28). RH in the enclosure was fixed at $65 \pm 5\%$ with a saturated salt solution of ammonium nitrate (NH_4NO_3) which is considered as the optimal %RH to favor carbonation [31]. The CO_2 curing system was placed in the room at 20 °C. CO_2 feeding was not continuous but the gas was injected regularly at given intervals of time with the regulator on the CO_2 tank. The CO_2 curing process was conducted in the following manner:

- A partial vacuum was created in the enclosure in order to reach an absolute pressure $P_{Vacuum} = 0.50 \pm 0.05$ bar.
- CO_2 was injected until the absolute pressure reached atmospheric one (about 1 bar).

This ensures to have an enclosure with $[CO_2] = 50\%$ v/v just after the CO_2 injection in standard temperature conditions and atmospheric pressure.

According to Šavija and Luković [34], cyclic CO_2 exposure results in a significantly increased degree of carbonation compared to continuous carbonation by preventing saturation due to released water that quickly fills the small pores as carbonation proceeds. Using cyclic carbonation (as is the case here) is finally an interesting method to re-establish pathways for CO_2 diffusion (part of pores are periodically emptied).

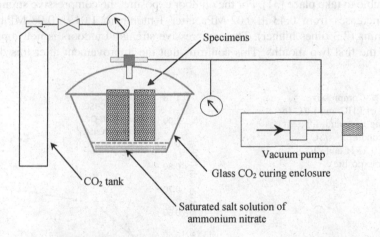

Fig. 4.28 Illustration of the CO_2 curing system

4.4.2   Mechanical Performances

4.4.2.1   Natural Carbonation (ISC/OC)

The compressive strength of bio-based concretes cured under natural carbonation for increasing ages is reported in Fig. 4.29.

The first observation concerns the lower compressive strength of LRC compared to LHC regardless of curing conditions and age. This was already identified at early ages in the previous part. By following the strength evolution until 10 months, a ceiling effect appears for LRC. The strength seems to be limited between 4 months and 10 months whereas it continues to increase quite significantly for LHC. As a result, the strength gain over time for LHC is much greater. Its compressive strength after 10 months under standard curing conditions (ISC) is 0.73 ± 0.03 MPa whereas that of LRC is only 0.38 ± 0.04 MPa. Another important result is the difference between indoor and outdoor curing beyond 2 months. The strength gain is higher for specimens exposed outdoors whether for LHC or LRC. This highlights a sharp improvement of the carbonation process in the conditions that occurred during the outdoor exposure. As shown before (Fig. 4.26b), RH mostly ranged between 45 and 75% during outdoor exposure curing and carbonation kinetics is known to be maximal between 50% and 70%RH. When RH is higher than 70%, pores tend to saturate with water, making the CO_2 diffusion through the concrete very slow. On the other hand, when RH is lower than 50%, pores tend to become dry and the dissolution of $Ca(OH)_2$ and CO_2 necessary for the carbonation reaction has trouble to take place [31]. For the outdoor exposure, the compressive strength of LHC increases from 0.43 ± 0.02 MPa after 1 month to 1.01 ± 0.08 MPa after 10 months (2.3 times higher). The compressive strength outdoors is not improved during the first two months. This confirms that the improvement after this date is

Fig. 4.29 Compressive strength of LHC and LRC for increasing ages and different curing conditions (*ISC* 20 °C–50%RH and *OC* Outdoor exposure)

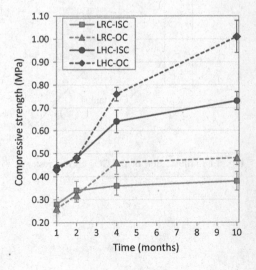

mostly due to enhanced carbonation. Indeed, at 50%RH, carbonation is expected to begin quite significantly only from 30 days according to the drying rate of plant-based concretes reported in Fig. 4.27. On the other hand, the higher RH outdoors until 2 months has a great chance to delay the beginning of carbonation by hindering CO_2 diffusion in the saturated pores at the early age.

It was shown that C_2S hydration kinetics at 50%RH is very slow. However, a significant acceleration of C_2S hydration was noted on NHL3.5/CL90–S mortars cured at 95%RH. As a consequence, the contribution of C_2S hydration could contribute to the strength evolution over time, especially for the outdoor exposure.

The higher compressive strength of specimens cured outdoors is necessarily linked to the hardening kinetics of the binder. However, the difference between LRC and LHC can also be explained by the lower bond strength between rice husks and lime. The inter-particles porosity and the mechanical response of the granular skeleton should not serve to justify the lower strength gain of LRC over time.

The trend observed on compressive strength are similar for elastic moduli with a more pronounced ceiling effect for LRC compared to LHC and a higher modulus for specimens exposed outdoors.

The age and the curing conditions of plant-based concretes prove to be closely linked to their hardening kinetics and their strength development over time. Consequently, the evolution of the chemical nature of the lime-based binder phases over time will be investigated in order to compare the hardening (i.e., carbonation and hydration) of the binder in both plant-based concretes.

4.4.2.2 Accelerated Carbonation Curing (ACC)

The compressive strength measured after 40 days of initial curing (ISC) and 30 days of accelerated carbonation curing (ACC) is compared with that achieved for the specimens cured 2 or 10 months under natural conditions in Fig. 4.30.

The results show that the compressive strength after ACC is approximately equivalent to that obtained after 10 months of outdoor exposure. The compressive

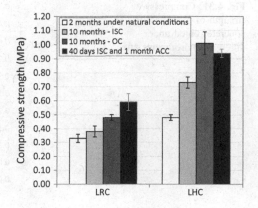

Fig. 4.30 Compressive strength of plant-based concretes after ACC compared to natural conditions

strength of LHC almost reaches 1 MPa obtained after 10 months outdoors and LRC exhibits a compressive strength of 0.59 ± 0.06 MPa which is even higher than that reached after 10 months outdoors. Furthermore, the compressive strength after ACC was doubled if compared to that measured after 2 months under natural conditions (i.e., natural carbonation).

The trend is the same for cycle moduli. They are higher than those obtained after 10 months of outdoor exposure for LRC and almost identical for LHC.

4.4.2.3 Moist Curing (MC) and Thermal Activation (TA)

These curing conditions are the same as those studied on lime-based mortars in the third chapter of this book (presented in Table 3.7). However, the drying kinetics of plant-based concretes is different from that of lime-based mortars. About 25 days are necessary to reach constant weight at 20 °C and 50%RH (ISC). Before being tested under compression, water-saturated specimens (MC or TA) were placed in a drying oven at 50 °C until their mass has reached that of samples cured 28 days under ISC (~ 48 h). This step is important to prevent water from disrupting the measurement of mechanical performances. The average bulk density of specimens before the compression test was 641 ± 13 kg m^{-3} for LRC and 456 ± 12 kg m^{-3} for LHC. Results are presented in Fig. 4.31.

The compressive strength of LRC remains the same regardless of curing conditions (~ 0.28 MPa). Elastic moduli follow a similar trend with a slight decrease especially for 21d–MC samples (Table 4.8). Furthermore, a sharp decrease in the compressive strength of LHC is noted for 7d–MC (30%) and 21d–MC/7d–TA samples (50%) in comparison to standard curing (28d–ISC). The same observation can be made with the modulus of LHC which decreased by 25% for 7d–TA samples and 35% for 21d–MC samples.

For both concretes, results are completely different from those obtained for lime-based mortars in the third chapter (Fig. 3.11). Mechanical performances even

Fig. 4.31 Compressive strength of plant-based concretes cured under different environments after 28 days

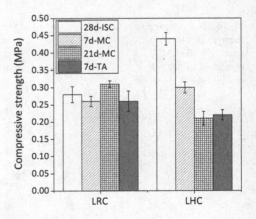

Table 4.8 Average modulus (E_C) of plant-based concretes at 28 days

Concrete	28d–ISC	7d–MC	28d–MC	7d–TA
LRC	50 ± 4	49 ± 2	38 ± 5	46 ± 2
LHC	73 ± 2	68 ± 2	48 ± 5	57 ± 3

follow an opposite trend for LHC since moist curing and elevated temperature have resulted in a dramatic fall of compressive strength. As regards LHC, the results are in accordance with the study of Arnaud and Gourlay [4]. The authors demonstrated that a humid environment (98%RH) slows down very sharply the setting of hemp concrete even when it is manufactured with hydraulic lime-based binders. It should be noted that LRC is less sensitive to curing conditions since the mechanical performances can be considered as almost unchanged.

In light of these results, the influence of curing conditions on the binder hardening and the physico-chemical interaction between the binder and the aggregates have to be investigated.

4.4.3 Binder Hardening

4.4.3.1 Under Natural and Accelerated Carbonation

Rate of Carbonation

The rate of carbonation (ROC%) can be first estimated measuring the weight gain of specimens after curing. Indeed, carbonation results in an increase of mass as molar mass of $CaCO_3$ is 35.1% higher than that of $Ca(OH)_2$. This method can cause errors associated with the high production of water during ACC since 1 mol of H_2O is released for each mol of CO_2 consumed by lime. It was decided to store the specimens 8 additional days after ACC in the room at 20 °C and 50%RH. Specimens were weighed after hydric stabilization and the weight was compared with that measured just before ACC and after 40 days of pre-conditioning in the same environment to overcome this problem of released water. Furthermore, it makes sense to follow the bulk density over time for specimens cured under ISC to have an indication of the carbonation process. However, for specimens exposed outdoors, it is erroneous to refer to the weight gain to follow the carbonation process since plant-based concrete are hygroscopic materials with an unstable density due to moisture uptake and release. Bulk densities of specimens cured under natural conditions (ISC/OC) for increasing ages or before and after ACC are presented in Table 4.9.

The weight gain of specimens in relation to the $Ca(OH)_2$ content [$W_G/Ca(OH)_2$] is as follows (4.13):

Table 4.9 Bulk density of specimens for increasing ages in kg m^{-3}

Time	Curing	LRC	LHC
1 month	ISC	625 ± 2	454 ± 5
	OC	666 ± 4	687 ± 5
10 months	ISC	654 ± 3	476 ± 2
	OC	489 ± 4	503 ± 3
40 days	ISC	638 ± 8	462 ± 3
After ACC	ACC	702 ± 13	509 ± 5

$$W_G/Ca(OH)_2(\%) = \frac{\Delta M}{m[Ca(OH)_2]} \times 100 \qquad (4.13)$$

where ΔM is the weight gain of specimens (with 8 additional days at 20 °C–50% RH for ACC) and $m[Ca(OH)_2]$ is the mass of calcium hydroxide used for the concrete manufacturing.

The rate of carbonation (ROC%) was defined as the conversion rate of $Ca(OH)_2$ into $CaCO_3$. It was calculated as follows (4.14):

$$ROC(\%) = \frac{W_G/Ca(OH)_2(\%)}{35.1} \times 100 \qquad (4.14)$$

The percentage increase in the bulk density of specimens cured under ISC up to 10 months is about 5%, giving a ROC of 33% after 10 months for both plant-based concretes. As regards ACC, the increase in the bulk density is 10% for LRC and LHC, thus giving a ROC of 72%. It must be mentioned that the water released by the carbonation process could be partially used for C_2S hydration. This can cause a minor error in ROC estimation by weight gain.

The phenolphthalein spray test was used to provide information about the carbonation depth. The solution was sprayed on the specimen section just after the compression test. The indicator half-way stage occurs for a pH which is about 9. The region appears unstained when pH is under 9 (totally carbonated area) whereas it is stained in pink when pH is beyond 9 (partially carbonated area). Carbonation profiles obtained by the phenolphthalein spray test after 10 months under natural conditions (ISC–10 m or OC–10 m) and after ACC are reported in Fig. 4.32.

For plant-based concretes cured at 20 °C and 50%RH (ISC–10 m), the section is almost entirely stained in pink without any difference between LRC and LHC. It does not mean that lime is uncarbonated. As was explained by Lawrence et al. [35], a stained area indicates a transitional state between the start and the finish of the carbonation process. If pH is not low enough (higher than 9), the area remains stained. However, a carbonation front is visible for specimens cured outdoors (OC–10 m). The carbonation depth according to the colorless region is about 0.8 cm for LRC and 1.5 cm for LHC. This confirms that carbonation has been promoted in outdoor conditions as previously assumed in view of strength development (Fig. 4.29).

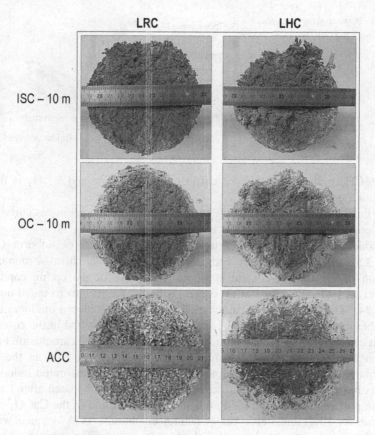

Fig. 4.32 Cross sectional view of specimens after failure in compression a few seconds after spraying with phenolphthalein

After ACC, the carbonation depth is approximately the same than that obtained after 10 months of outdoor exposure. Nevertheless, the core of LRC specimens appears more carbonated than that of LHC as the bulk area is stained in pale pink (Fig. 4.32).

In addition, TGA was used to investigate the carbonation kinetics of the lime-based binder. After the compression test, powdered matrix samples were collected in the bulk of the specimens (LRC-B and LHC-B) and on their surface (LRC-S and LHC-S) (Fig. 4.33).

As it was done for lime-based mortars (Table 3.4), $Ca(OH)_2$ and $CaCO_3$ contents were determined and the rate of carbonation (ROC) was calculated as follows (4.15):

$$ROC(\%) = \frac{\%CH_0 - \%CH_T}{\%CH_0} \times 100 \qquad (4.15)$$

Fig. 4.33 Powder sampling and sieving for TGA

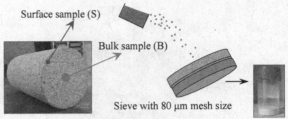

Surface sample (S)

Bulk sample (B)

Sieve with 80 μm mesh size

Binder powder for TGA

where $\%CH_0$ is the initial CH content in unhydrated lime and $\%CH_T$ is the CH content at a given date.

The time evolution of the $CaCO_3$ content in the core of specimens cured under natural conditions up to 10 months is firstly reported in Fig. 4.34.

Considering that lime powder initially contains 8–10% of unburnt $CaCO_3$ (Table 3.5), the core is slightly carbonated after 1 month. Up to 4 months, the evolution of the $CaCO_3$ content is similar regardless of the curing conditions. However, at 10 months, the $CaCO_3$ content is higher for specimens cured outdoors (Fig. 4.34). The higher rate of carbonation of specimens exposed outdoors is only noticeable in the long term by TGA since the binder is collected in the core. Once again, it can be concluded that outdoor exposure has promoted carbonation because of better RH conditions which enhanced the dissolution of CO_2 in the pores. Furthermore, TGA on surface samples collected on specimens cured indoors and outdoors has revealed that the surface is almost entirely carbonated after 1 month.

An important result from Fig. 4.34 is the same evolution of the $CaCO_3$ content over time for LRC and LHC. Under natural conditions, TGA coupled with the phenolphthalein spray test shows that the higher compressive strength of LHC is not due to a better carbonation kinetics within LHC. The rate of carbonation in the

Fig. 4.34 $\%CaCO_3$ in bulk samples (LRC–B and LHC–B) of specimens up to 10 months under natural conditions and after ACC

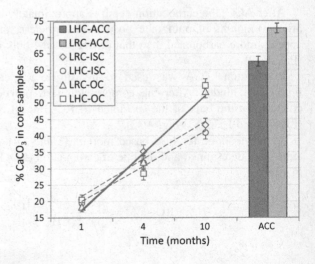

Table 4.10 Overview of ROC for specimens cured 10 months under natural conditions (ISC/OC) and after accelerated carbonation (ACC) depending on the method (weight gain and TGA)

Curing	Time	Concrete	Weight gain	TGA
			ROC (%)	
ISC	10 months	LRC	32.5	42.6
		LHC	33.6	39.6
OC	10 months	LRC	–	55.6
		LHC	–	57.9
ACC	After ACC	LRC	72.1	80.4
		LHC	71.5	67.3

core of specimens obtained after 10 months of natural curing is about 40% indoors and slightly less than 60% outdoors (Table 4.10). It is very close for LRC and LHC. Only the carbonation depth is somewhat bigger for LHC (Fig. 4.32).

The $CaCO_3$ content in the bulk of specimens after ACC is secondly reported in Fig. 4.34. It is higher in the core of LRC. This result is in accordance with the carbonation profiles obtained by the phenolphthalein spray test (see Fig. 4.32). The rates of carbonation obtained by TGA in the core of specimens are compared with those obtained by weight gain in Table 4.10. After 10 months of natural curing or after ACC, the rate of carbonation by weight gain is the same for both plant-based concretes. In fact, this method provides an overall assessment of the carbonation process from the surface to the core of the specimen whereas TGA gives a local measurement in the core. For this reason, TGA highlights the different response of plant-based concretes with regard to the reactivity of CO_2 under ACC. In addition, this is confirmed by the phenolphthalein test. For LHC, the rate of carbonation in the core after ACC is rather close to that obtained after 10 months outdoors (67% after ACC and 58% after 10 months outdoors) whereas the rate of carbonation in the core of LRC after ACC is significantly higher (80%). The accelerated carbonation implies changes in the binder microstructure in such a way that CO_2 reactivity-diffusivity mechanisms differ for LRC and LHC under accelerated conditions. LRC becomes more favorable to CO_2 diffusivity in the core with a high CO_2 concentration. Excess CO_2 which has not already reacted with available hydrated lime penetrates deeper into the specimen. The higher inter-particles porosity within LRC (Fig. 4.18) probably plays a role in this enhanced CO_2 diffusivity under accelerated conditions. For LHC, its local reactivity is better given that the unstained region is bigger. This can be explained by the lower amount of lime to carbonate (see mix proportions in Table 4.7.) or by a barrier effect since CO_2 diffusivity throughout the totally carbonated area appears to have been hindered.

Hydration

The weight loss occurring between 100 and 400 °C by TGA is due to the loss of water from C–S–H hydrates [36]. TGA curves in this range of temperature are reported in Fig. 4.35.

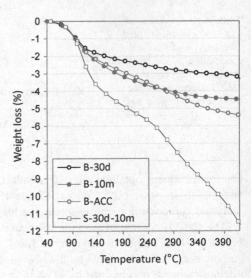

Fig. 4.35 TGA curves for lime samples collected in plant-based concretes between 40 and 400 °C (*B* Bulk, *S* Surface)

Firstly, it must be mentioned that the loss of water in bulk samples is the same for LHC and LRC regardless of the age of the concrete and the curing conditions (B–30d, B–10 m, B–ACC). However, it can be noted that the hydration rate has increased between 30 days and 10 months to the same extent for LHC and LRC. Furthermore, the C–S–H content in surface samples is higher than in the bulk as of 30 days (S–30d–10 m). It has been previously demonstrated that the carbonation rate has considerably increased in core samples from 30 days to 10 months and that the surface was quickly carbonated after 30 days. Therefore, a synergy effect between carbonation and hydration is assumed since the hydration rate is higher when the carbonation of hydrated lime is more advanced.

The correlation between carbonation and hydration is represented in Fig. 4.36. The loss of water between 100 °C and 400 °C (H_2O bound to C–S–H) and the CO_2 weight loss are reported for bulk samples collected in specimens under natural conditions (ISC and OC) from 1 to 10 months and after ACC. The surface samples are not considered in the graph because the weight loss in the range 250–400 °C is suspected to be partially attributed to the decomposition of cellulosic compounds. The evolution of the C–S–H content correlates fairly well with the carbonation rate with a regression coefficient of about 0.9. This linear correlation shows that C_2S hydration is promoted by $Ca(OH)_2$ carbonation. This can be explained by the water locally provided by the carbonation reaction which could benefit C_2S hydration, meaning that the initial mixing water is not effective to perform C_2S hydration at early ages. In the section about lime-based binders, it has been seen that C_2S hydration strongly depends on curing conditions. With relatively dry conditions (50%RH indoors—under 80%RH outdoors—65%RH under ACC), C_2S hydration after 1 month is very low. In these conditions, it appears that carbonation prevails over C₂S hydration. In the longer term, when carbonation of hydrated lime is sufficiently advanced, C_2S hydration is promoted by the water provided by carbonation. Based on this assumption and considering that the linear correlation is linked to the stoichiometry of the carbonation reaction, the slope (=7.8) leads to a molar ratio $n(H_2O)/n(CO_2) \approx 0.3$. According to this approach, about 30% of the water released by carbonation could be used for C_2S hydration at the most.

4.4.3.2   Under Moist Curing and Thermal Activation

Rate of Carbonation

Arnaud and Gourlay [4] explained in part the decrease in the compressive strength of hemp concrete under moist conditions (as evidenced in Fig. 4.31) by the water-blocking of pores thus limiting CO_2 diffusion and carbonation. However, the average carbonation rate in the bulk of plant-based concretes after 28 days is about 13% regardless of curing conditions (28d–ISC, 7d–MC, 21d–MC or 7d–TA). In fact, the drying time of plant-based concretes is a little less than 1 month (Fig. 4.5). In addition, it was shown that CO_2 diffusion in the bulk of LHC/LRC specimens can only occur beyond 15 days (Fig. 4.27). Consequently, the carbonation rate is low in every case and mainly due to carbonic acid transfer in liquid phase. The beginning of air hardening will be probably delayed after 28 days for plant-based concretes cured under moist conditions. Furthermore, it was shown that the compressive strength of CL90–S mortars cured under moist conditions has been only marginally affected (21d–MC) or not at all (7d–MC and 7d–TA) (Fig. 2.11). Therefore, the carbonation process is not a determining factor for the early age mechanical strength of bio-based concretes obtained under moist curing and thermal activation.

Fig. 4.36 Correlation between %CO₂ release by decarbonation of CaCO₃ and loss of water bound to C–S–H hydrates for bulk samples under natural conditions and after ACC

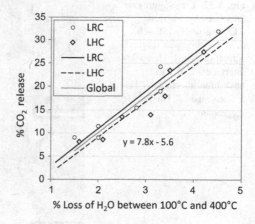

Hydration

TGA shows that both moist-cured and thermally activated samples are characterized by a stronger weight loss in the range 200–300 °C, compared to standard ones (28d–ISC). Moist curing and elevated temperature lead to a higher extraction of carbohydrate polysaccharides from the plant aggregates towards the binder. This can be explained by the water-soaked state in which aggregates remain during the curing period. This probably leads to a stronger decomposition of xylans which are easily removable in alkaline pore water.

Transition Zone Between Lime and Plant Aggregates

SEM pictures of plant aggregate-lime interfaces are reported in Fig. 4.37 for ISC and MC/TA since similar results are obtained for moist-cured and thermally activated samples. As regards hemp-lime interface under ISC (Fig. 4.37a), a thin gap of about 5 μm is observed between the particle and the matrix. For MC/TA samples, the gap zone turns into a gaping hole which is more than 200 μm thick (Fig. 4.37b). The hemp particle is totally uncoupled from the binder, suggesting a strong lack of adhesion. Moreover, the particle is surrounded by a porous matrix (up to 2 mm). Both debonding gap and affected binder surrounding the aggregates correspond to the interfacial transition zone (ITZ) which is obviously of poor quality in the case of moist curing and thermal activation.

The observations on hemp-lime interface can be explained by capillary pressure and moisture transport between aggregates and lime. After mixing, the presence of a water film and a binder zone with a higher water content surrounding water-saturated hemp aggregates is assumed (Fig. 4.38a).

Fig. 4.37 Lime-aggregate interfaces by SEM. **a, c** Curing under ISC—**b, d** Moist curing or thermal activation—**a, b** Hemp-lime interface—**c, d** Rice husk-lime interface. *H* Hemp, *R* Rice husk, *B* Binder, *Ag* Aggregate

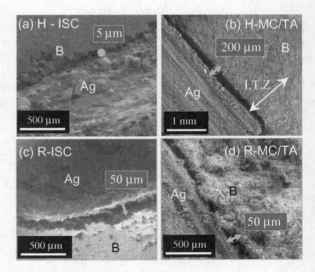

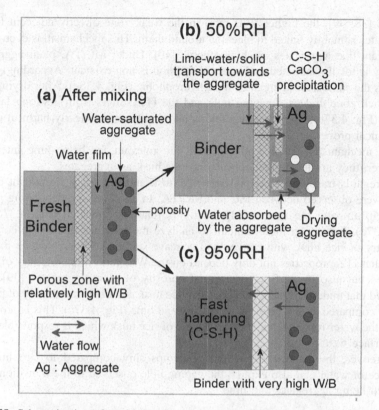

Fig. 4.38 Schematic view of mechanisms involved in the influence of curing conditions on the interface between the lime binder and hemp aggregates. **a** After mixing. **b** Under ISC. **c** Under MC

Under standard conditions (ISC: 20 °C–50%RH), wet aggregates begin to dry just after demolding and a certain amount of water is transported into hurd channels by capillary absorption. As a result of this initial flow rate from the binder to the hemp aggregates, a physical transport of lime-water or solid particles towards the aggregate is assumed. These movements can result in precipitation of hydration products and $CaCO_3$ near the aggregate surface and even within the particle. Under ISC, it is assumed that the interfacial zone changes from a water-filled zone to a zone increasingly filled with solids and whose porosity reduces with time (Fig. 4.38b) [37].

Under MC/TA, hemp shives remain water-saturated and hydration kinetics of the binder is faster than that performed under ISC. Partial hydration of the binder results in a finer pore structure and water consumption. Thus, water is probably squeezed out of hemp channels by capillary pressure as these plant aggregates are characterized by coarse pores. This will inevitably lead to excess water around aggregates during all the curing period (7 or 21 days). This remaining water film surrounding the aggregates involves a locally high W/B ratio in the ITZ and provides a higher porosity of the binder than in the bulk paste. According to some authors [38, 39], the higher W/B ratio in the weak zone directly adjacent to the aggregates is mainly linked to bleeding around them. This mechanism is even more significant that aggregates are hydrophobic [10]. Under MC/TA, plant aggregate porosity is totally saturated and water absorption is non-existent. Accordingly, the result is the same as if aggregates were hydrophobic (Fig. 4.38c). After drying, the water rich zone creates a gaping hole and the ITZ suffers from a strong lack of binder (Fig. 4.37b). These unfavorable bond conditions are clearly harmful to the mechanical properties of LHC.

The mechanisms detailed in Fig. 4.38 are relevant for hemp-lime interface. However, they are less suited considering rice husk as aggregates.

As regards transition zone between rice husk and lime, any significant differences were observed between ISC and MC/TA. In all cases, the debonding gap is about 50 μm and the surrounding matrix is relatively affected (Fig. 4.37c and Fig. 4.37d). It is much more difficult to analyze the interface due to the complex geometry of rice husk with convex and concave surfaces. Nonetheless, it may be noted that ITZ properties not only depend on the W/B ratio in the vicinity of plant particles but also on surface texture and porosity of plant aggregates [39]. It is assumed that under ISC, adhesion between rice husk and lime was already of poorer quality compared to that between hemp shiv and lime (Fig. 4.37c). This is probably due to the water-repellent and smooth cuticle of rice husk which is responsible for a low surface wettability.

Moreover, the important swelling of hemps shiv compared to rice husk in presence of water can also explain the gaping hole observed in the transition zone between hemp aggregates and lime.

4.4.4   Conclusion About Curing Regime and Mechanical Performances of Plant-Based Concretes

- The results of previous investigations have shown that the hardening kinetics of the lime-based binder in LHC and LRC is equivalent under natural conditions (20 °C–50%RH or outdoor exposure). After 10 months of curing, carbonation and hydration rates of plant-based concretes are very close. The hardening is not more disturbed for LRC. However, the gradual hardening of LRC provides only a very low strength gain over time compared to LHC. In light of these findings, the lower strength of LRC is necessarily explained by the weaker bond strength between rice husk and lime and the granular stacking of LRC (with the adverse effect of the high inter-particles porosity). Nevertheless, the granular stacking cannot justify the fact that the strength gain of LHC over time is twice as high as that of LRC. The effect of moist curing on the mechanical strength of bio-based concretes after 1 month confirmed that the bond strength between rice husk and lime is of poorer quality regardless of RH in the environment. The unfavorable granular stacking of LRC is certainly responsible for the lower mechanical

strength at early age (irrespective of the binder hardening over time and the bond strength of the binder with the particle) whereas the weaker bond strength between rice husk and lime leads to a limited strength gain of LRC over time.
- The accelerated carbonation curing involves a different response of plant-based concretes with regard to CO_2 diffusivity and reactivity. As a result, the cyclic CO_2 curing is found to be efficient for increasing the carbonation rate in the core of LRC at early ages (80%). However, the overall carbonation rate of plant-based concretes after ACC is equivalent.
- When RH is in the range 50–70%, the carbonation process prevails over C_2S hydration which is very slow. Even with a rather low content of C_2S, the hardening of lime-based mortars is significantly accelerated when RH is over 95% (owing to increased hydration rate). However, the interfacial transition zone between the aggregate and the lime-based binder is affected by moist curing due to a strong leaching of polysaccharides in the binder and excess water in the vicinity of plant aggregates, especially in the case of LHC.
- The best way to accelerate the hardening and the strength development of plant-based concretes is to establish optimal curing conditions for carbonation (i.e., 65%RH and cyclic CO_2 curing).

4.5   Studying the Shear Behavior of Plant-Based Concretes

4.5.1   Interest of the Analysis of the Mechanical Behavior of Plant-Based Concretes Under Shear Loading

Plant-based concretes are only considered as insulating materials whether they are cast in situ or in the form of prefabricated blocks. As a matter of fact, the structural design practice of wood frame walls associated with LHC does not assume any contribution of the plant-based material. In view of their properties, it makes sense to consider that plant-based concretes could contribute to the mechanical perfor-mance of the structure. In particular, some authors [23, 40] have shown that LHC provides in-plane racking strength to the timber frame. According to Munoz and Pipet [40], the mechanical behavior of a timber stud frame with LHC infill was enhanced compared to that with diagonal bracing. The studwork frame with LHC exhibited higher stiffness, racking strength and strain capacity at failure. Gross and Walker [5] studied the racking strength of a timber studwork encapsulated with low density LHC (320 kg m^{-3}). These authors concluded that even with a low strength (compressive strength was about 0.4 MPa after 5 months), manually tamped LHC improves the racking performance of timber studwork frames. An illustration of the racking strength test and failure of timber wall is presented in Fig. 4.39.

Mukherjee [23] has found that hemp-lime prevents weak axis buckling of timber columns by acting as a continuous lateral elastic support. Regarding high density LHC (715 kg m^{-3}), it was stated that the latter could add strength to the wall by

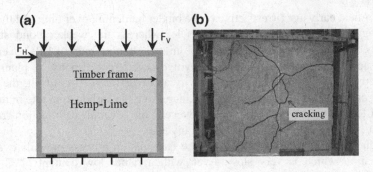

Fig. 4.39 a Illustration of a racking strength test on timber frame with LHC infill, F_H *Horizontal* load, F_V *Vertical* load. **b** Cracking of the panel according to Gross and Walker [5]

partly contributing to its load-bearing capacity. High density was performed by increasing the binder content in this study.

These works show that timber sections could be reduced and some design practice of timber frame panels should be reviewed without noggins and diagonal braces. In this context, more knowledge on the shear behavior of plant-based concrete is necessary to optimize the structural design.

4.5.2 Experimental Results for Triaxial Compression on LHC and LRC

4.5.2.1 Deviatoric Responses

Figure 4.40 displays the evolution in the deviatoric stress as a function of the axial strain with rising confining pressures (from 25 to 150 kPa) for LHC. An increase in the peak deviatoric stress with increasing confining pressure is noted. For $p'_0 = 25$ kPa, the maximum deviatoric stress is around 2.7 MPa whereas it is 3.3 MPa for $p'_0 = 150$ kPa. Moreover, for the vast majority of the specimens, one observes a clear evolution towards a behavior with a stronger ductility when the confining pressure increases. This strain hardening is especially noticeable for a confining pressure of 150 kPa. In this case, the strain capacity at peak stress reaches 19%. This trend has already been observed for ordinary concrete and cemented sands [41, 42]. However, for high confining pressures, some specimens exhibit a more brittle behavior (lower strain and higher elastic modulus than the other ones). These specimens (marked with a hollow circle on the curves reported in Fig. 4.40) show a failure mode with a localized shear band (noted FM1 in Fig. 4.40). After the initiation of the shear plane, a reduction in the peak deviatoric stress occurs quite suddenly. This well-defined localized failure with shear banding is known to be responsible for an accelerated softening response at post-peak strength [42]. Nevertheless, this kind of failure only concerns a very small number of LHC specimens. Actually, most of them rather show localized bulging and crushing in their lower part (noted FM2 in Fig. 4.40). The shear failure surface is thus less distinct or even invisible. For all specimens that show this failure mode, high confining pressures involve gradual deformation of the specimens and high ductility. The failure patterns of LHC can be observed on specimens in Fig. 4.42a, b.

The deviatoric stress is plotted against the axial strain for LRC specimens in Fig. 4.41.

As for LHC, the peak deviatoric stress increases with increasing confining pressure, but to a lesser extent (from about 1.5 to 1.75 MPa). The loading capacity of LRC under shearing becomes sensitive to confinement for $p'_0 = 100$ kPa. A similar comment applies to ductility which is especially improved when the confining pressure is 150 kPa. A major and interesting outcome of the post-test observations is the failure mode of LRC which corresponds to shear banding for all specimens except for one (over the 12 specimens tested). The associated failure pattern is represented in Fig. 4.42c. The consistency of this failure mode (FM1) for LRC is higher than that of the bulging mode (FM2) observed on LHC. Among LRC specimens, the only one that exhibits bulging (Fig. 4.42d) shows an important strain hardening between 3 and 18% strain before reaching a plateau (Fig. 4.41). This level of ductility is not achieved for other specimens.

The Young's modulus (E_C) is plotted against the confining pressure in Fig. 4.43. Since it proves to be sensitive to the failure mode of specimens, the only way to study the effect of the confining pressure on E_C is to consider its values on a case-by-case basis. As expected, the modulus of the LHC specimens for which shear banding (FM1) clearly occurs is significantly higher than that reported for the other ones. In addition, some LHC specimens for which shear banding is not necessarily visible also exhibit a high modulus (for $p'_0 = 25$ and 50 kPa). By contrast, when pure bulging occurs (FM2), a particularly low modulus is measured. Between these extremes, an intermediate modulus corresponds to a failure that

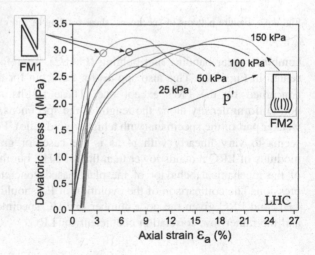

Fig. 4.40 Deviatoric stress versus axial strain with growing confining pressure for LHC with two kinds of failure modes (FM1 and FM2)

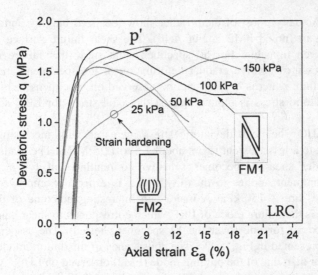

Fig. 4.41 Deviatoric stress versus axial strain with growing confining pressure for LRC

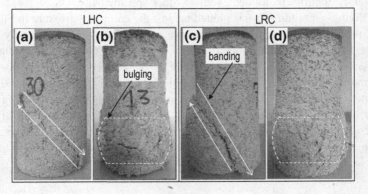

Fig. 4.42 Failure patterns of specimens: shear banding (**a** and **c**), bulging failure (**b** and **d**)

combines shear banding and bulging (FM1/2). Therefore, three groups are represented in Fig. 4.43. This also applies for LRC but for the latter, all specimens are concerned by the FM1 except one. These results show that in the case of non-uniform density along the length of the specimens, the modulus is that of the weaker part of the specimen (with a higher void ratio). For a given failure mode, E_C seems to vary linearly with p' as is the case for granular materials [43]. The modulus of LRC remains lower than that of LHC but this is not surprising in view of the mechanical behavior of the plant-based concretes under unconfined compression. The comparison of the evolution of E_C should not be performed between LHC and LRC given the poor number of LHC specimens concerned by the FM1 and the rather low correlation coefficient for LRC.

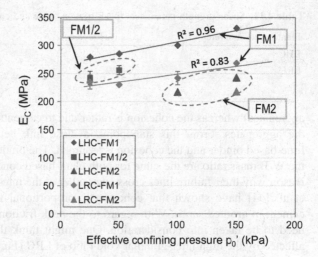

Fig. 4.43 Effect of initial effective confining pressure on Young's modulus

4.5.2.2   Evaluation of Shear Strength Parameters in ($q - \sigma'_m$)

Mean values of the peak deviatoric stress (q_P) for increased mean effective pressure are plotted in Fig. 4.44.

The failure lines of bio-based concretes under shearing are given by a linear regression of the peak values (q_P, σ'_m). Thereafter, the peak friction angle (φ_P) and the cohesion (C) are determined with (4.10) and (4.11) exposed before and reported in Table 4.11.

Firstly, the peak friction angle of LHC (46°) is found to be higher than that of LRC (29°). However, both plant-based concretes show a same cohesion which is about 0.36 MPa. Based on the model used for ordinary concrete [41] or even for sands and clays reinforced by cement grouting [44, 45], the friction angle is more willingly related to the granular skeleton (that is to say the interlocking of the plant

Fig. 4.44 Peak shear strength of bio-based concretes in ($q - \sigma'_m$)

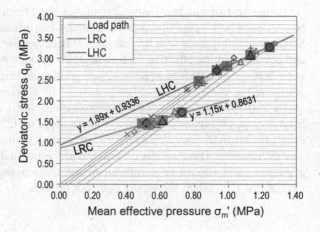

Table 4.11 Shear strength parameters (peak friction angle and cohesion)

Plant-based concrete	φ_P (degrees)	C (kPa)
LHC	46	355
LRC	29	362

aggregates) whereas the cohesion is rather due to cementation and bonding between the aggregates. From this standpoint, a close link between the strength of the lime-based binder and the cohesion is assumed. The binder content (B/A = 2.3) and the W/B mass ratio are the same for both plant-based concretes. This is probably the reason why their failure lines cross each other at the intercept. For instance, Maalej et al. [44] have shown that cohesion is proportional to the volume fraction of cement in grouted sands. With regard to the peak friction angle, several parameters need to be taken into consideration. One might think that the latter is negatively affected by the high inter-granular void ratio of LRC (Fig. 4.17). Moreover, particle size distribution has also a great influence on the friction angle. The particle size range of hemp shives is assumed to provide a better packing of the particles thus contributing to the shear strength of LHC. The very different shape of the plant particles (thin semi-ellipsoidal husk vs. thick parallelepipedic shiv), their rigidity and their surface properties (as roughness) are also inherent features of aggregates which could benefit to the friction angle of LHC.

The friction angle of plant-based concretes is mobilized only to the extent that the normal stress is significant whereas the contribution of cohesion to shear strength is always available. If the wall section is not exposed to horizontal forces and if the vertical load is negligible, a possible conclusion that would be drawn is that the cohesion strength (0.36 MPa) could be taken as a safe shear strength for design.

Furthermore, the different shear behavior of the plant-based concretes is likely to depend upon their anisotropy. It is possible that the stratified arrangement of LHC, presumed to be more prominent [6, 7], may have contributed to the bulging failure mode compared to LRC which is more isotropic.

To the best of our knowledge, no studies have been made about the frictional properties of bulk hemp shives and rice husks. For instance, the study of Aloufi and Santamarina [46] deals with the mechanical behavior of rice grains. However, some authors reported the effective friction angle of woodchips measured with the direct shear box test [47, 48]. Morphology, size grading and rigidity of woodchips can be considered as close to those of hemp shives. The difference lies in the bulk density (from 200 to 300 kg m^{-3} regarding woodchips [48]) and the internal porosity. Stasiak et al. [47] report an effective friction angle of 42° whereas the values of Wu et al. [48] are in the range of 44° to 54° depending on moisture content and particle size distribution. These results correspond to the internal friction angle of unbound aggregates and should be compared to the friction angle at large strains in the triaxial test (i.e., when the binder does not provide any more cohesion).

4.5.3 First Conclusions Regarding the Triaxial Compression of Plant-Based Concretes

Under triaxial compression, the ductility of specimens increases with the effective confining pressure. However, ductility and elastic modulus prove to be strongly impacted by the failure mode of specimens under the shear loading. In most cases, the failure mode of LHC is a combination of bulging localized in the lower part of the specimens and shear banding. When banding prevails on bulging, the mechanical behavior of LHC is less ductile and the elastic modulus is much higher. LRC exhibits a higher consistency of the failure mode corresponding to a clean shear banding in virtually all specimens. Therefore, it is assumed that bulging of LHC is mainly due to the non-uniform pore distribution along the specimens (explained by the vibro-compaction process in a single layer) but also to its shear behavior which is assumed to be more anisotropic than that of LRC.

Plant-based concretes show a same cohesion (0.36 MPa) but the peak friction angle of LHC (46°) is higher than that estimated for LRC (29°). The cohesion seems to be highly correlated with the binder strength. Furthermore, this study highlights the predominant influence of the aggregate type on the peak friction angle. Inherent features of plant aggregates (size, shape, roughness, stiffness), inter-granular void ratio and manner that particles are packed probably all have a certain role to play in achieving lower shear strength for LRC. The strain capacity of plant-based concretes is such that specimens do not reach the critical state at large strains in the present work.

References

1. C. en Chanvre, *Constuire en Chanvre. Règles professionnelles d'éxécution* (SEBTP. Société d'Édition du Bâtiment et des Travaux Publics, 2012)
2. S. Amziane, L. Arnaud, *Les bétons de granulats d'origine végétale. Application au béton de chanvre* (Lavoisier, France, 2013)
3. M. Chabannes, *Formulation et étude des propriétés mécaniques d'agrobétons légers isolants à base de balles de riz et de chènevotte pour l'éco-construction* (University of Montpellier, France, 2015), p. 215
4. L. Arnaud, E. Gourlay, Experimental study of parameters influencing mechanical properties of hemp concretes. Constr. Build. Mater. **28**(1), 50–56 (2012)
5. C. Gross, P. Walker, Racking performance of timber studwork and hemp-lime walling. Constr. Build. Mater. **66**, 429–435 (2014)
6. P. Tronet, T. Lecompte, V. Picandet, C. Baley, Study of lime hemp composite precasting by compaction of fresh mix—An instrumented die to measure friction and stress state. Powder Technol. **258**, 285–296 (2014)
7. P. Tronet, T. Lecompte, V. Picandet, C. Baylet, Study of lime and hemp concrete (lhc)—Mix design, casting process and mechanical behaviors. Cem. Concr. Compos. **67**, 60–72 (2016)
8. T.M. Dinh, Contribution au développement du béton de chanvre préfabriqué utilisant un liant pouzzolanique innovant, Ph.D. Thesis, Toulouse 3 University (Paul Sabatier), France, p. 211, 2014

9. A. Evrard, Transient hygrothermal behaviour of Lime-Hemp Materials, Ph.D. Thesis, Catholic University of Louvain, Belgium, p. 140, 2008
10. V. Nozahic, Vers une nouvelle démarche de conception des bétons végétaux lignocellu-losiques basée sur la compréhension et l'amélioration de l'interface Liant/Végétal. Application à des granulats de chènevotte et de tige de tournesol associés à un liant ponce/chaux, Ph.D. Thesis, Clermont University, France, p. 311, 2012
11. T.T. Nguyen, Contribution à l'étude de la formulation et du procédé de fabrication d'éléments de construction en béton de chanvre, Ph.D. Thesis, Bretagne-Sud University, France, p. 167, 2010
12. N. Yüksel, The review of some commonly used methods and techniques to measure the thermal conductivity of insulation materials, in *Insulation Materials in Context of Sustainability* (InTech, 2016), pp. 113–140
13. M. Li, H. Zhang, Y. Ju, Design and construction of a guarded hot plate apparatus operating down to liquid nitrogen temperature. Rev. Sci. Instrum. **83**(7) (2012)
14. F. Collet, S. Pretot, Thermal conductivity of hemp concretes: variation with formulation, density and water content. Constr. Build. Mater. **65**, 612–619 (2014)
15. R. Coquard, D. Baillis, D. Quenard, Experimental and theoretical study of the hot-wire method applied to low-density thermal insulators. Int. J. Heat Mass Transf. **49**(23–24), 4511–4524 (2006)
16. V. Cerezo, Propriétés mécaniques , thermiques et acoustiques d'un matériau à base de particules végétales : approche expérimentale et modélisation théorique, Ph.D. Thesis, École Nationale des Travaux Publics de l'État, Lyon, France, p. 243, 2005
17. J. Chamoin, Optimisation des propriétés (physiques, mécaniques et hydriques) de bétons de chanvre par la maîtrise de la formulation, Ph.D. Thesis, Rennes 1 University, INSA Rennes, France, p. 198, 2013
18. M. Ghafghazi, D. Shuttle, Confidence and accuracy in determination of the critical state friction angle. Soils Found. **49**(3), 391–395 (2009)
19. F. Becquart, F. Bernard, N.E. Abriak, R. Zentar, Monotonic aspects of the mechanical behaviour of bottom ash from municipal solid waste incineration and its potential use for road construction. Waste Manag **29**(4), 1320–1329 (2009)
20. P. Glé, E. Gourdon, L. Arnaud, Acoustical properties of materials made of vegetable particles with several scales of porosity. Appl. Acoust. **72**(5), 249–259 (2011)
21. F. Collet, M. Bart, L. Serres, J. Miriel, Porous structure and water vapour sorption of hemp-based materials. Constr. Build. Mater. **22**(6), 1271–1280 (2008)
22. S. Elfordy, F. Lucas, F. Tancret, Y. Scudeller, L. Goudet, Mechanical and thermal properties of lime and hemp concrete ('hempcrete') manufactured by a projection process. Constr. Build. Mater. **22**(10), 2116–2123 (2008)
23. A. Mukherjee, C. MacDougall, Structural benefits of hempcrete infill in timber stud walls. Int. J. Sustain. Build. Technol. Urban Dev. **4**(4), 295–305 (2013)
24. D. Sedan, Etude des interactions physico-chimiques aux interfaces fibres de chanvre/ciment. Influence sur les propriétés mécaniques du composite, Groupe d'étude des Matériaux Hétérogènes, Ph.D. Thesis, Limoges University, France, p. 129, 2007
25. R. Walker, S. Pavía, Effect of Hemp'S soluble components on the physical properties of Hemp concrete. J. Mater. Sci. Res. **3**(3), 12–23 (2014)
26. Y. Diquélou, E. Gourlay, L. Arnaud, B. Kurek, Impact of hemp shiv on cement setting and hardening: influence of the extracted components from the aggregates and study of the interfaces with the inorganic matrix. Cem. Concr. Compos. **55**, 112–121 (2014)
27. V.D. Pizzol, L.M. Mendes, L. Frezzatti, H. Savastano, G.H.D. Tonoli, Effect of accelerated carbonation on the microstructure and physical properties of hybrid fiber-cement composites. Miner. Eng. **59**, 101–106 (2014)
28. V.D. Pizzol, L.M. Mendes, H. Savastano, M. Frías, F.J. Davila, M.A. Cincotto, V.M. John, G. H.D. Tonoli, Mineralogical and microstructural changes promoted by accelerated carbonation and ageing cycles of hybrid fiber–cement composites. Constr. Build. Mater. **68**, 750–756 (2014)

29. J. Lanas, J.L.P. Bernal, M. Bello, J.I. Galindo, Mechanical properties of natural hydraulic lime-based mortars. Cem. Concr. Res. **34**, 2191–2201 (2004)
30. S. Asavapisit, G. Fowler, C. Cheeseman, Solution chemistry during cement hydration in the presence of metal hydroxide wastes. Cem. Concr. Res. **27**(8), 1249–1260 (1997)
31. P. Turcry, L. Oksri-Nelfia, A. Younsi, A. Aït-Mokhtar, Analysis of an accelerated carbonation test with severe preconditioning. Cem. Concr. Res. **57**, 70–78 (2014)
32. V. Wiktor, F. De Leo, C. Urzì, R. Guyonnet, P. Grosseau, E. Garcia-Diaz, Accelerated laboratory test to study fungal biodeterioration of cementitious matrix. Int. Biodeterior. Biodegradation **63**(8), 1061–1065 (2009)
33. T. Colinart, P. Glouannec, P. Chauvelon, Influence of the setting process and the formulation on the drying of hemp concrete. Constr. Build. Mater. **30**, 372–380 (2012)
34. B. Šavija, M. Luković, Carbonation of cement paste: understanding, challenges, and opportunities. Constr. Build. Mater. **117**, 285–301 (2016)
35. R.M.H. Lawrence, T.J. Mays, P. Walker, D. D'Ayala, Determination of carbonation profiles in non-hydraulic lime mortars using thermogravimetric analysis. Thermochim. Acta **444**(2), 179–189 (2006)
36. S. Xu, J. Wang, Y. Sun, Effect of water binder ratio on the early hydration of natural hydraulic lime. Mater. Struct. **48**(10), 3431–3441 (2014)
37. J.C. Maso, Interfacial transition zone in concrete. RILEM Report 11, London, 1996
38. K.O. Kjellsen, R.J. Detwiler, O.E. Gjørv, Pore structure of plain cement pastes hydrated at different temperatures. Cem. Concr. Res. **20**(6), 927–933 (1990)
39. J.-K. Kim, Y.-H. Moon, S.-H. Eo, Compressive strength development of concrete with different curing time and temperature. Cem. Concr. Res. **28**(12), 1761–1773 (1998)
40. P. Munoz, D. Pipet, in *Bio-aggregate-Based Building Materials: Applications to Hemp Concretes*, eds. S. Amziane, L. Arnaud. Plant-based Concretes in Structures: Structural Aspect-addition of a Wooden Support to Absorb the Strain (WILEY-ISTE., 2013)
41. L. Zingg, M. Briffaut, J. Baroth, Y. Malecot, Influence of cement matrix porosity on the triaxial behaviour of concrete. Cem. Concr. Res. **80**, 52–59 (2016)
42. A. Marri, The mechanical behaviour of cemented granular materials at high pressures, Ph.D. Thesis of University of Nottingham, UK, p. 279, 2010
43. F. Becquart, Caractérisation du comportement mécanique d'un mâchefer dans la perspective d'une méthodologie de dimensionnement adaptée aux structures de chaussées," in *XXIVe Rencontres Universitaires de Génie Civil*, Nantes, France, 2006
44. Y. Maalej, L. Dormieux, J. Canou, J.C. Dupla, Strength of a granular medium reinforced by cement grouting. Comptes Rendus Mec. **335**(2), 87–92 (2007)
45. S. Horpibulsuk, Mechanism controlling undrained shear characteristics of induced cemented clays. Lowl. Technol. Int. **7**(2), 9–18 (2005)
46. M. Aloufi, J.C. Santamarina, Low and high strain macrobehavior of grain masses—the effect of particle eccentricity. Food Proc Eng Inst ASAE **38**, 877–887 (1995)
47. M. Stasiak, M. Molenda, M. Bańda, E. Gondek, Mechanical properties of sawdust and woodchips. Fuel **159**, October 2016, 900–908 (2015)
48. M.R. Wu, D.L. Schott, G. Lodewijks, Physical properties of solid biomass. Biomass Bioenerg. **35**(5), 2093–2105 (2011)
49. C. Magniont, Contribution à la formulation et à la caractérisation d'un écomatériau de construction à base d'agroressources, Ph.D. Thesis, Toulouse III University—Paul Sabatier, France, p. 343, 2010
50. Z. Pehlivanli, I. Uzun, Z.P. Yücel, I. Demir, The effect of different fiber reinforcement on the thermal and mechanical properties of autoclaved aerated concrete. Constr. Build. Mater. **112**, 325–330 (2016)
51. Z. Pehlivanli, İ. Uzun, İ. Demir, Mechanical and microstructural features of autoclaved aerated concrete reinforced with autoclaved polypropylene, carbon, basalt and glass fiber. Constr. Build. Mater. **96**, 428–433 (2015)

52. M. Albayrak, A. Yörükoğlu, S. Karahan, S. Atlihan, H. Yilmaz Aruntaş, I. Girgin, Influence of zeolite additive on properties of autoclaved aerated concrete. Build. Environ. **42**(9), 3161–3165 (2007)
53. M. Jerman, M. Keppert, J. Výborný, R. Černý, Hygric, thermal and durability properties of autoclaved aerated concrete. Constr. Build. Mater. **41**, 352–359 (2013)
54. E.P. Aigbomian, M. Fan, Development of Wood-Crete building materials from sawdust and waste paper. Constr. Build. Mater. **40**, 361–366 (2013)
55. J. Wu, G. Bai, H. Zhao, X. Li, Mechanical and thermal tests of an innovative environment-friendly hollow block as self-insulation wall materials. Constr. Build. Mater. **93**, 342–349 (2015)
56. M. Sutcu, J.J. Del Coz Diaz, F.P. Alvarez Rabanal, O. Gencel, S. Akkurt, Thermal performance optimization of hollow clay bricks made up of paper waste, Energy Build. **75**, 96–108 (2014)
57. K.S. Shibib, H.I. Qatta, M.S. Hamza, Enhancement in thermal and mechanical properties of bricks. Therm. Sci. **17**(4), 1119–1123 (2013)

Chapter 5
Conclusion and Outlooks

The results reported in this document first contribute to the development of an innovative plant-based concrete using raw rice husk. This crop residue is available throughout the year at low cost and its use to manufacture bio-based concretes for green building provides a new recovery sector for a local by-product coming from rice farming in France.

The different origin of rice husk compared to plant particles coming from stalks (hemp shives, sunflower aggregates, etc.) means that physical (density, water absorption) and morphological properties of rice husk are very different.

Hemp shiv and rice husk were mixed with lime-based binders and two different casting processes were used. The first one corresponds to manual tamping as done by workers on the building site and the second one is based on vibro-compaction of the freshly-mixed plant-based material. For a given binder-to-aggregate mass ratio and manual tamping, it is almost impossible to reach a same apparent density for hemp and rice-husk based concretes owing to the apparent density of rice husk which is more than twice that of hemp shiv. With the vibro-compaction process, it was possible to achieve a close density for plant-based concretes by increasing the density of hemp concrete through the reduction of the macroscopic inter-particles porosity.

The inter-particles porosity of lime and rice husk concrete is considerably higher than that of hemp-lime concrete. This high amount of voids is such that the rice-husk based concrete can compete with hemp concrete in terms of thermal conductivity. However, the adverse effect of this granular stacking partly explains the lower mechanical performances of the concrete using rice husk as aggregate, irrespective of curing conditions.

Plant-based concretes cast by manual tamping have shown a same hardening kinetics of the lime binder over 10 months under natural carbonation (20 °C, 50% RH or outdoor exposure) while the strength development of rice husk concrete over time was strongly limited compared to that of hemp concrete. This was attributed to the weaker bond strength of the binder with rice husk.

© The Author(s) 2018
M. Chabannes et al., *Lime Hemp and Rice Husk-Based Concretes for Building Envelopes*, Biobased Polymers, https://doi.org/10.1007/978-3-319-67660-9_5

The accelerated carbonation curing (cyclic CO_2 exposure) at 20 °C and 65%RH was found to be effective to increment the short-term compressive strength of plant-based concretes. Therefore, it could be used as part of a block making factory.

Moist curing (95%RH) led to a strong increase in the compressive strength of lime-based mortars including C_2S (like hydraulic lime). The hardening of lime-based binders can be significantly accelerated at early ages. Results showed that high RH and elevated temperature (50 °C here) promoted C_2S hydration. Nevertheless, these conditions were counterproductive for bio-based concretes since the interfacial transition zone between the particles and the binder was affected by moist curing due to excess water in the vicinity of plant aggregates. Hence, the best way to accelerate the hardening of plant-based concretes is to establish optimal curing conditions for carbonation.

Some investigations about the shear behavior of plant-based concretes by means of triaxial compression are pioneers and have made possible the determination of the shear strength parameters. A consistent value of cohesion was achieved and attributed to the binder strength while a different peak friction angle was related to the aggregate contribution. The shear strength of plant-based concrete was found to be significant. Consequently, it should be considered for the design practice of building envelopes.

Here are some outlooks for the research presented in the book:

- The compressive strength of rice husk concrete is lower than that of hemp concrete even if a same mix proportioning is used. The packing of rice husks could be enhanced by increasing the rice husk content in the concrete while reducing the binder content for trying to reduce the inter-particles porosity without increasing the density of the concrete and the environmental impact. It appears challenging to achieve a higher compressive strength than that reported in this research with vibro-compaction but it might be a good opportunity to increase ductility and shear strength. Another way to increase the packing density of lime and rice husk concrete would consist in adding rice straw in the mix.
- For both plant-based concretes, a more accurate approach is required to predict the shear contribution of bio-based concrete in relation to their mix proportioning.
- Further studies are needed to investigate the shear strength of low density plant-based concretes cast by manual tamping for monolithic construction. Moreover, the size effect of specimens should be considered. A large-scale triaxial test would be more representative of the real conditions.
- There has been an increasing interest in producing blocks at an industrial scale. These are self-supporting and typically inserted into wood frames [1]. The effect of curing conditions (and especially the CO_2-curing) should be explored on vibro-compacted specimens. The cumulative effect of these two processes could significantly increase the short term mechanical performances of industrial blocks (under compression and shear loading).

- Under natural conditions (outdoor curing) or accelerated carbonation, a further comprehensive study about the coupling effect of relative humidity, CO_2 content and ventilation should be considered to provide optimal curing conditions.
- In the case of precast blocks with high compaction of the mixture at the fresh state, thermal and mechanical anisotropies of hemp concrete have already been proven [2]. When the block is inserted in such a way that the vertical load in the wall is parallel to the compaction direction, mechanical strength is increased but thermal conductivity tends to be less favorable. When the block is rotated by 90° (the load is perpendicular to the compaction direction), it is the opposite. The shear strength of the block in this configuration has every chance to be far below that measured in this research. Attention will have to be paid to these aspects including the case of rice husk concrete.

References

1. A. Arrigoni, R. Pelosato, P. Melià, G. Ruggieri, S. Sabbadini, G. Dotelli, Life cycle assessment of natural building materials: The role of carbonation, mixture components and transport in the environmental impacts of hempcrete blocks. J. Clean. Prod. **149**, 1051–1061 (2017)
2. V. Nozahic, Vers une nouvelle démarche de conception des bétons végétaux lignocellulosiques basée sur la compréhension et l'amélioration de l'interface Liant/Végétal. Application à des granulats de chènevotte et de tige de tournesol associés à un liant ponce/chaux, Ph.D. Thesis (Clermont University, France, 2012), p. 311

Reproduction of figures—Reference list

Figure II-1c. International Journal of Biological Macromolecules, vol. 17, no. 6, M.R. Vignon, C. Garcia-Jaldon, D. Dupeyre, Steam explosion of woody hemp chenevotte, pages No. 395–404, 1995, with permission from Elsevier.

Figure II-4. Chemical Society Reviews, vol. 41, no. 24, D.M. Alonso, S.G. Wettstein, J.A. Dumesic, Bimetallic catalysts for upgrading of biomass to fuels and chemicals, pages No. 8075–8098, 2012—Reproduced by permission of the Royal Society of Chemistry.

Figure II-7a. Construction and Building Materials, vol. 14, no. 8, R. Jauberthie, F. Rendell, Origin of the pozzolanic effect of rice husks, pages No. 419–423, 2000, with permission from Elsevier.

Figure II-7b, c, d. Biomass and Bioenergy, vol. 25, no. 3, B.-D. Park, S.G. Wi, K.H. Lee, A.P Singh, T.-H. Yoon, Y.S. Kim, Characterization of anatomical features and silica distribution in rice husk using microscopic and micro-analytical techniques, pages No. 319–327, 2003, with permission from Elsevier.

Figures II-6, II-9, II-12, IV-3, IV-12 and IV-20. Construction and Building Materials, vol. 70, M. Chabannes, J.-C. Bénézet, L. Clerc, E. Garcia-Diaz, Use of raw rice husk as natural aggregate, an innovative application, pages No. 428–438, 2014, with permission from Elsevier.

Figures III-3 and III-4. Journal of Thermal Analysis and Calorimetry, Quantitative study of hydration of C_3S and C_2S by thermal analysis, vol. 102, no. 3, 2010, pages No. 965–973, S. Goni, F. Puertas, M. Soledad Hernandez, M. Palacios, A. Guerrero, J.S. Dolado, B. Zanga, F. Baroni, with permission of Springer.

Figures III-5, III-6, III-11, IV-31, IV-37 and IV-38. Construction and Building Materials, vol. 102, M. Chabannes, E. Garcia-Diaz, L. Clerc, J.-C. Bénézet, Effect of curing conditions and $Ca(OH)_2$-treated aggregates on mechanical properties of rice husk and hemp concretes using a lime-based binder, pages No. 821–833, 2016, with permission from Elsevier.

Figure III-7. Thermochimica Acta, vol. 444, no. 2, R.M.H. Lawrence, T.J. Mays, P. Walker, D. D'Ayala, Determination of carbonation profiles in non-hydraulic lime mortars using thermogravimetric analysis, pages No. 179–189, 2006, with permission from Elsevier.

© The Author(s) 2018
M. Chabannes et al., *Lime Hemp and Rice Husk-Based Concretes for Building Envelopes*, Biobased Polymers, https://doi.org/10.1007/978-3-319-67660-9

Figure III-9. Cement and Concrete Research, vol. 23, no. 4, C. Shi, R.L. Day, Acceleration of strength gain of lime-pozzolan cements by thermal activation, pages No. 824–832, 1993, with permission from Elsevier.

Figure III-10. Cement and Concrete Research, vol. 34, J. Lanas, J.L.P. Bernal, M. Bello, J. I. Galindo, Mechanical properties of natural hydraulic lime-based mortars, pages No. 2191–2201, 2004, with permission from Elsevier.

Figures IV-14, IV-15, IV-40, IV-41, IV-42, IV-43 and IV-44. Construction and Building Materials, vol. 143, M. Chabannes, F. Becquart, E. Garcia-Diaz, N-E. Abriak, L. Clerc, Experimental investigation of the shear behaviour of hemp and rice husk-based concretes using triaxial compression, pages No. 621–632, 2017, with permission from Elsevier.

Figures IV-23 and IV-24. Journal of Materials Science Research, vol. 3, no. 3, R. Walker, S. Pavia, Effect of hemp's soluble components on the physical properties of hemp concrete, pages No. 12–23, 2014, with permission from JMSR.

Figures IV-26, IV-27, IV-28, IV-29, IV-30, IV-32, IV-33, IV-34 and IV-36. Construction and Building Materials, vol. 94, M. Chabannes, E. Garcia-Diaz, L. Clerc, J-C. Bénézet, Studying the hardening and mechanical performances of rice husk and hemp-based building materials cured under natural and accelerated carbonation, pages No. 105–115, 2015, with permission from Elsevier.

Figure IV-39. Construction and Building Materials, vol. 66, C. Gross, P. Walker, Racking performance of timber studwork and hemp-lime walling, pages No. 429–435, 2014, with permission from Elsevier.

Printed in the United States
By Bookmasters